Deploying RFID - Challenges, Solutions, and Open Issues

Deploying RFID - Challenges, Solutions, and Open Issues

Editor

Edvin Falk

scitus academics

Deploying RFID - Challenges, Solutions, and Open Issues
Edited by **Edvin Falk**

ISBN: 978-1-68117-289-7
Library of Congress Control Number: 2016937465

SCITUS Academics LLC,
www.scitusacademics.com
Box No. 4766, 616 Corporate Way,
Suite 2, Valley Cottage,
NY 10989

Printed in United States of America on Acid Free Paper

Preface

RFID refers to radio frequency identification. The acronym means small electronic devices made up of an antenna and a small chip. The chip typically contains 2,000 bytes or less data. Radio frequency identification (RFID) is an increasingly popular technology for a wide range of applications, such as inventory tracking, supply chain management, automated manufacturing, healthcare, etc. The benefits of RFID technology can be seen in terms of efficiency or effectiveness (customer services provided by companies). Radio Frequency Identification (RFID) automatically identifies and tracks tags used by electromagnetic fields attached to objects. The tags contain electronically stored information. Passive tags gather energy from a nearby RFID reader's interrogating radio waves. The RFID device serves the same purpose as a bar code or a magnetic strip on the back of a credit card or ATM card; it provides a unique identifier for this object. And, just as a bar code or a magnetic strip must be scanned for information, the RFID device must be scanned for data retrieval. RFID tags, a technology used to be restricted to tracking cattle, track consumer products worldwide. Many manufacturers use the tags to track the location of each product until it is removed from the shelf and placed in a shopping cart. An important advantage of RFID devices over the above is that the RFID device does not need to be positioned precisely in relation to the scanner. We all know how difficult it is sometimes for shopkeepers to read a bar code. Credit cards and ATM cards clearly have to be swiped by a special reader. RFID technology has been available for over fifty years. The ability to produce RFID devices has only recently fallen to the point where they can be used as an inventory or control device. Some common problems with RFID are readers and tags colliding. A clash with the reader occurs when two or more readers' signals overlap. The tag cannot answer requests simultaneously.

This book will be of great importance to engineers, researchers and industry staff and will give them new ideas to tackle the challenges of today and tomorrow.

Table of Contents

Chapter 1 Impacts of RFID on Business Models 1

Chapter 2 The Role of RFID Technology in Supply Chain Risk Management 17

Chapter 3 RFID-Based Vehicle Positioning and Its Applications in Connected Vehicles 49

Chapter 4 Investigation of Interference Models for RFID Systems 71

Chapter 5 RFID-Based Equipment Monitoring System 101

Chapter 6 A UHF/UWB hybrid silicon RFID tag with on–chip antennas 119

Chapter 7 Using RFID Technology for Simplification of Retail Processes 143

Chapter 8 A Computational Architecture Based on RFID Sensors for Traceability in Smart Cities 173

Chapter 9 Creating TUIs Using RFID Sensors—A Case Study Based on the Literacy Process of Children with Down Syndrome 227

Chapter 10 A systematic review of RFID applications and diffusion: key areas and public policy issues 255

Index 289

CHAPTER 1

Impacts of RFID on Business Models

Ela Sibel Bayrak Meydanoğlu[1]

[1] Marmara University, Turkey

1. INTRODUCTION

Business model describes the business logic a company or network of companies use to generate revenue and create customer and network value. It enables to identify how inputs of a company or network of companies are transformed to value-adding outputs (Kamoun, 2008). RFID affects business models through guiding new business models (e.g. RFID-enabled pay-per-use business model) or reshaping existing ones and thereby enables value creation. The number of studies that deals with the mentioned impact of RFID on business models is limited. Understanding the impact of RFID on business models is essential for companies to create a business value by using RFID in order to gain competitive advantage. This study aims to illustrate and clarify the mentioned impact. Thereby it provides an important contribution to the limited studies in the relevant literature.

In this study initially the term "business model" is defined. Subsequently the business model framework adopted in the study is presented and the major components of RFID systems are reviewed briefly. This is followed by brief explanations about RFID business models.

Furthermore it is discussed how RFID systems influence the components of the adopted business model framework and how they contribute to reshape existing business models or create new ones. The study concludes with the main findings and implications.

2. METHODOLOGY

This conceptual study is a basic research, which is executed based on the previous studies about business models, business model frameworks, benefits of RFID systems as well as the limited studies about the impact of RFID systems on the components of business model frameworks. Its aim is to reorganize the existent ideas in order to give an insight how companies can create business value from RFID technology.

3. BUSINESS MODELS

3.1. Definition of business models

Different definitions for the term business model and its building blocks exit in the relevant literature. Table 1 below includes the definitions of some authors.

Based on the definitions in Table 1 business model can be defined as a model that describes how a company or network of companies creates value from new products, innovations, activities for business partners and customers.

3.2. Business model framework

Basic building blocks of a business model and the external forces that have an affect on these blocks are described in a business model framework. In the relevant literature various business model frameworks are proposed (e.g. framework of Kamoun (Kamoun, 2008), framework of Shi and Manning (Shi & Manning, 2009), framework of Osterwalder, Pigneur and Tucci (Osterwalder et al., 2005)). Among these frameworks the value-driven framework of Kamoun (see Figure 1), which is constructed considering the inherent use of business model in conceptualizing the value creation and money earning logic of a company or network of companies, is adopted in this study.

The business model framework of Kamoun encompasses four main components: value proposition, value creation system, value deliverance and value capture model. Each of these four components is further divided into subcomponents (see Table 2).

Table 1. Business Model Definitions

Author(s)	Definition
Timmers (1998), (as cited in Schweizer, 2005)	Timmers defines business model as *"an architecture for the product, service and information flows including a description of the various business actors and their roles, a description of the potential benefits for the various business actors and a description of the sources of revenues."*
Hamel (2000), (as cited in Schweizer, 2005)	Hamel defines four elements that form a business model: customer interface, core strategy, strategic resources and value network. The customer interface and the value network represent the relation between buyer and supplier side. The core strategy represents the mission of the company and the scope of production. Strategic resources explain competitive advantage gained through competencies and assets deployed. Competencies and assets support the underlying strategy through customer benefits resulting from the core strategy and through company boundaries intermediating between the strategic resources and the value network where a company is positioned.
Hoppe and Kollmer (2001), (as cited in Schweizer, 2005)	Hoppe and Kollmer define business model as an integrated and consistent picture of a company that illustrates the way it aims to generate revenues.
Magretta (2002), (as cited in Schweizer, 2005)	Magretta defines business model as a story that explains how companies work and that contains motivation and a plan that describes how value is delivered.
Betz (2002)	*"A business model is an abstraction of a business identifying how that business profitably makes money. Business models are abstracts about how inputs to an organization are transformed to value-adding outputs."*

Osterwalder, Pigneur and Tucci (2005)	*"A business model is a conceptual tool that contains a set of elements and their relationships and allows expressing the business logic of a specific firm. It is a description of the value a company offers to one or several segments of customers and of the architecture of the firm and its network of partners for creating, marketing and delivering this value and relationship capital to generate profitable and sustainable revenue streams."*
Kamoun (2008)	Kamoun defines business model as *"the logic a company or network of companies use to generate revenue and create customer and network value."* According to Kamoun business model can be defined as a blueprint that defines the way a business creates and captures value from new services, products or innovations.
Shi and Manning (2009)	Shi and Manning define business model as *"the outcome of management actions – planned, emergent or realized – in defining a firms's offerings and activities."*

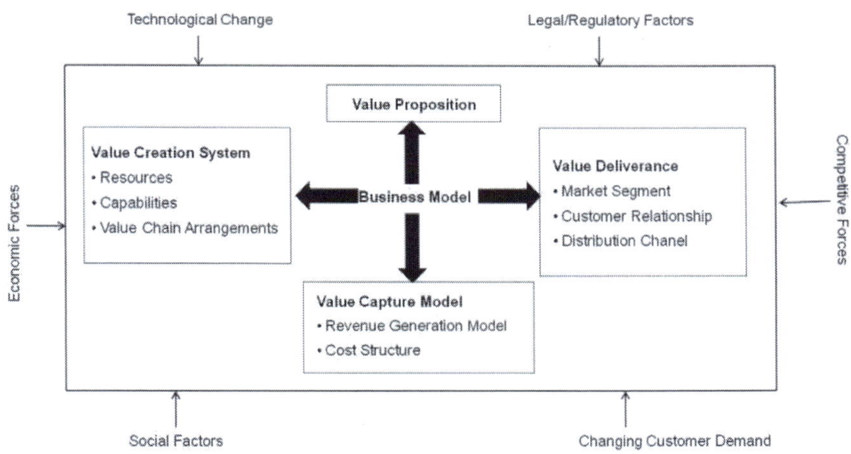

Figure 1. Business Model Framework of Kamoun (Kamoun, 2008)

As shown in Figure 1 the components of the defined framework are influenced by external environmental forces "economy", "technological change", "legal/regulatory factors", "social factors", "competitive forces" and "customer demand".

Table 2. Components of Business Model Framework of Kamoun (Kamoun, 2008)

Component	Description
Value Proposition	This component defines the added value that is offered by a company or network of companies to its customers via a bundle of new products, services or innovations.
Value Creation System	This component consists of subcomponents that are necessary to create the above mentioned added value.
Resources	Financial, physical, human, technological and organizational resources that are necessary to execute the business model and deliver the proposed added value.
Capabilities	Skills that are necessary to coordinate the resources.
Value Chain Arrangement	With this component the structure of the value chain is intended. Under the structure the inter-linked activities and alliances of the company with suppliers, partners and distributors have to be understood. These activities and alliances are necessary to execute the business model and deliver the proposed value.
Value Deliverance	This component consists of subcomponents that are relevant to deliver the proposed value.
Market Segment	This component represents the group of customers and geographic markets a company wants to deliver the proposed value. Different segments might have different needs. As a result of this, different products, services and value proposition might be required.
Customer Relationship	This component represents the link the company establishes with its customer to deliver the proposed value.
Distribution Channel	This component represents the way a company transmits its customers its products, services that have an added value for customers.
Value Capture Model	This component consists of subcomponents that are relevant to revenue and costs that arise from the proposed value.
Revenue Generation Model	This component defines how the income is generated. It defines the impact of the proposed value on revenue.
Cost Structure	This component defines the costs that arise to execute the business model and deliver the proposed value.

4. RFID-SYSTEMS

RFID is an Auto-ID technology that enables to identify tagged items by means of radio waves. Main components of a RFID system are:

- *Tag (Transponder)*: It consists of an antenna and a microchip. Microchip stores data about the tagged item. Antenna transmits the data about the tagged item to the reader by means of radio waves (Kavas, 2007).

- *Reader (Transceiver)*: It is a device that communicates with tags through radio waves and reads data on them (Karygiannis et al., 2007).
- *RFID Middleware*: It is a software that is used to consolidate, aggregate, process and filter raw RFID data received from multiple readers to generate useful information for end-users. It transmits also the processed data to backend enterprise applications (Kamoun, 2008).
- *RFID System Software*: It is necessary for the communication between tags and readers. This software enables to read tags, write on tags, detect and fix erroneous data as well as to realize authentication for security (Üstündağ, 2008).
- *Backend Enterprise Service*: This service enables to receive filtered RFID data from the middleware and integrate these with existing applications such as ERP, SCM or CRM systems through Application Programming Interfaces (APIs) (Kamoun, 2008).

5. RESHAPING EXISTING BUSINESS MODELS OR CREATING NEW BUSINESS MODELS WITH RFID

RFID can reshape existing business models or create new ones. Instead of manual scanning of the products bought at paying counter, automatic scanning through RFID is an example for reshaping of a business model. Through reshaping, efficiency of business model can be increased. RFID is also used to reduce failure rate, shrinkage, operating stock and to enhance on-time delivery, shipment quality and so forth.

RFID technology can also be used to create a new business model that enables to generate new value creation opportunities and therewith to gain competitive advantage as well as to develop new ways to make money. Table 3 includes some examples for this type of RFID business models.

Table 3. Examples for RFID Business Models (Kamoun, 2008)

RFID Business Model	Description
RFID Infrastructure & Management Services Provider	Provider of RFID infrastructure-related products and solutions (e.g. tags, readers, data integration services, middleware)
RFID-enabled pay-per-use Business Model	According to this model a firm that has a huge number of assets in a given industry lets its trading partners to use these assets and to pay for the assets per use. The firm tracks its assets through RFID devices located throughout the supply chain.
RFID-based Security Provider	A firm that provides RFID-based tracking solutions for authentication, brand protection as well as to combat tampering, theft, counterfeiting.
Information and Business Intelligence Agent	A firm that uses RFID to offer new information-based services. It analyzes, for example with the use of data mining techniques, the extensive data collected from RFID systems and provides business intelligence solutions.

6. IMPACT OF RFID ON THE COMPONENTS OF THE ADOPTED BUSINESS MODEL FRAMEWORK

6.1. RFID and value proposition

RFID can provide a new or an added value for customers of companies that use this technology. As this value is a motive for customers to prefer the mentioned companies, it has a positive impact on the revenues of the companies. A new or an added value can be proposed by RFID:

- *Through creating a new way of conducting economic transactions among trading partners:* RFID enables to form new strategic networks such as buyer-supplier partnerships, pay-per-use business models. As RFID used in these networks provides various advantages (e.g. cost reduction, superior customer service level) for companies, they want to be a partner in such a network. For example, the tracking capability of RFID enables a company to realize pay-per-use business model that means invoicing the trading partner for the hired asset each time he uses the asset. Pay-per-use business models help trading partners, who participate in these models, to reduce purchasing, storage and maintenance costs as well as to eliminate losses due to stock thefts (Kamoun, 2008). For example, a furniture manufacturer can hire RFID-tagged sofas produced by him to a hotel according to pay-per-use model. He can monitor the usage of sofas with the help of RFID (e.g. a sofa can count the number of persons that sit on it, the person's weight and seating time) and create a monthly itemized billing statement to the hotel (Bohn et al., 2004). This model enables the hotel to exempt from purchasing and maintenance costs.

- *Through providing a superior customer service level:* Superior customer service level produced by using RFID increases customer satisfaction, which has a positive impact on the revenue stream of a company. For example, Metro aims to enrich the value of its product offerings through RFID-enabled smart shelves, smart dressing rooms that allow its customers to find the correct size, color and additional information about a displayed garment by touching a screen (Weber, 2003). Such a service can be a good motive for a consumer to go Metro for shopping.

- *Through reducing transaction costs by achieving transaction efficiencies:* RFID can increase the efficiency of executed transactions (Lin et al., 2006)). This causes reduction at transaction costs. Transaction efficiency and cost reduction can lead to lower prices that are an important motive for customers to prefer a company (Kamoun, 2008). For example, instead of scanning each product bought manually at the paying counter, a retailer can use RFID and scan automatically as well as instantaneously all bought products at the counter as the customer passes through a reader and exits the store. Thereby scanning and paying processes can be executed more quickly. This means time saving for customers and labor saving for companies that use RFID (Erickson & Kelly, 2007).

6.2. RFID and value deliverance

Among subcomponents of value deliverance RFID has an impact on customer relationship component. RFID can increase customer satisfaction that has a positive impact on revenue stream. For example in a store, which uses RFID, a customer can get information about the existence of a garment matching his style, size and color requirements and if the desired garment exist, the store's clerk can precisely locate the garment for him. If the garment is out of stock, customer can get information with the help of RFID-enabled system about the nearest store where the garment is available. Contactless checkouts are other examples that increase customer satisfaction by using RFID. This type of checkouts enables automatic scanning of RFID-tagged items in shopping carts by RFID readers at checkout counters. Readers enable also the automatic billing of customers using their RFID-tagged credit cards (Kamoun, 2008).

Gaining insights about consumer behaviours is essential to execute marketing activities successfully. Marketing activities aim to increase customer satisfaction and thereby to affect revenue stream positively. Data mining technologies enable to gain the necessary insights. It is possible to combine RFID with data mining technologies. Based on RFID-captured consumer behaviour valuable insights about consumer behaviour can be created with the help of data mining technologies. These insights are used to enhance responsiveness of companies to their customer preferences (Kamoun, 2008; Hoffmann et al., 2005). Clothing retailer, for example, can make promotional offers based on collected and analyzed information of buying habit (size, favorite colours etc.) of a consumer who has a loyalty card (Erickson & Kelly, 2007).

6.3. RFID and value capture

6.3.1. RFID and revenue model

As illustrated above RFID can increase customer satisfaction, strengthen customer loyalty and thereby has a positive impact on revenue (Kamoun, 2008).

RFID can also be used to protect manufacturers against counterfeits, which are a real threat to the revenues of manufacturers (Kamoun, 2008). Product authentication plays an important role to combat counterfeiting and to detect counterfeit products. It enables to determine whether a given product is genuine or counterfeit. RFID-based product authentication is an important technological measure for checking the originality of a product that moves in a network of companies. Up to consumer each actor in a network can be the entry point of the counterfeit product. To realize a secure network each actor in a network has to verify the authenticity of the products on hand. To understand whether a given product is genuine or counterfeit, the insertion of a security feature into the product and the authentication of this feature are essential. RFID tags can be used for the authentication of security features (Filimon, 2008).

RFID increases network visibility, which affects revenues positively (Kamoun, 2008; Erickson & Kelly, 2007; Lin et al., 2006). Out-of-stock situations, which means loss of revenues, can be prevented through a better visibility. Better network visibility prevents also to retain great amount of stock that can be sold at discounted prices, if it is not sold by the end of the season (Kamoun, 2008).

6.3.2. RFID and cost model

Cost structure is also an important subcomponent of the value capture component. Utilization of RFID in companies or company networks can contribute much to reduce costs. Below this contribution is illustrated based on some examples:

- RFID enables perpetual inventory that is important to get information about current inventory level. Through perpetual inventory time and costs for physical inventory are saved. Stocking more or less items as a result of false inventory information can also be prevented (LakeWest Group & MeadWestvaco Intelligent Systems, 2003).

- As every item from warehouses to distribution centers and from these centers to retail shelves can be tracked through RFID, a reduction in stockouts can be ascertained by using RFID. In companies or company networks items leaving shelves or facilities are automatically recorded via RFID, computers are updated in terms of existing stocks and purchasing is executed if inventory levels drop too low. With less worries about stockouts companies or networks hold less safety stocks. Reduction at stocks means less inventory costs (Erickson and Kelly, 2007; LakeWest Group & MeadWestvaco Intelligent Systems, 2003).

- As demand for stocking decreases it will be possible to use store floor, which is used before for stocking, for new products that enrich the product range of a company or network of companies. In other words additional merchandise will be available for sale without costly requirements of store design and remodel (LakeWest Group & MeadWestvaco Intelligent Systems, 2003).
- As mentioned above RFID enhances transaction efficiency that gives rise also to cost reductions. For example, automation of some processes (e.g. checkouts, incoming goods control) via RFID causes a reduction both at labor costs and costs that incur to rectify errors caused by manual execution (Erickson and Kelly, 2007).

6.4. RFID and value creation

In order to create value with RFID, resources such as IT personnel with proper training, RFID infrastructure (tags, readers, printers, antennas, computers, networking equipment, middleware, application software, integration software etc.) must exist. A budget must also be allocated (Kamoun, 2008).

To operate RFID systems and to create a value some capabilities are also necessary. Following some examples for capabilities are listed (Kamoun, 2008):

- Ability to deal with erroneous tag reads, data redundancy, damaged tags
- Ability to manage reader and tag collision, signal inference and noise
- Ability to overcome great amount data generated by readers to enhance existing knowledge base and enable intelligent decision making.

RFID affects the structure of value chains in three ways (Kamoun, 2008):

- It eliminates inefficiencies in existing value chains. For example, RFID enables to track the movements of goods from the store's back door to the point of purchase in real time. This leads to better inventory visibility.
- It can enhance the collaboration between chain partners. For example, the collaboration between retailers and suppliers can be enhanced through RFID that enables suppliers to optimize production and replenishment scheduling based on real-time demand.
- It can give rise to new strategic networks. RFID enabled pay-per-use business model is an example for such a strategic network.

6.5. RFID and external environmental factors

Among external factors shown in Figure 1 technological change, competitive forces and legal/regulatory factors are relevant for RFID business models.

As at the present day RFID infrastructure is costly the use of this technology is limited. However through technological developments it is expected that infrastructure costs will decrease. For example, it is expected that with the help of nano technology silicone chip demand will disappear and through ink based RFID circuits costs of tags will decrease (Kış, 2006).

Adversaries of RFID declare that RFID threatens data privacy. According to them through RFID consumers can be tracked and thereby not only the consumption behaviours of consumers but also their private lives and other habits can be tracked (LakeWest Group & MeadWestvaco Intelligent Systems, 2003). However through some methods data privacy can be protected. Using lock command, kill command, press-to-activate switch, blocker tags, clipped tags and electromagnetic shielding, active jamming, frequency hopping, encryption of data in transit, encryption of data stored on tags, authentication are some examples for these methods. Furthermore regulative countermeasures can play an important role to eliminate fears about data privacy. Garfinkel's manifesto - named RFID Bill of Rights - is an example for regulative countermeasures. According to this manifesto consumers that consume tagged products have the following rights (Korkmaz et al., 2006):

- To know whether a product is tagged
- To have a choice to accept, discard, disable or remove the tag
- To know which information is saved on tags
- To know when, where and why tags are read
- Not to lose their rights (e.g. right of product return) even if they prefer to buy products without tags or deactivate tags with kill command.

The firm IDTechEx projects a very rapid growth in RFID. A market research report, which is executed by this firm, contains the predictions in Table 4 about the deployment of RFID tags from 2005 to 2015 (Raafat et al., 2007). In an environment, in which the use of RFID increases rapidly, companies have to use RFID in order to outmatch and not to lose market share.

Table 4. Forecast for global sales of RFID tags in billions between 2005-2015 (Raafat et al., 2007)

Year Kategories in which tags are used	2005	2010	2015
Item	0.5	27.0	1,000.0
Pallet/Case	0.4	30.0	35.0
Other	0.4	5.7	12.5
All Kategories Total	1.3	62.7	1,047.5

7. CONCLUSION

In order to gain advantages over competitors, companies or network of companies can use technological innovations to reshape their business models or to create new ones. RFID is one of these innovations. This technology reshapes business models through automating transactions (e.g. contactless checkouts). Thereby it contributes to increase transactional efficiency. RFID can also be used to create a new business model through developing new ways to make money. For example, a firm that offers business intelligence solutions with the use of data mining techniques can integrate its business intelligence software with RFID and start to analyze data collected from RFID systems. Thereby the firm can offer its customers a new information-based service that brings it money. RFID users can propose an added value or a new value for their customers. Added or new value increases customer satisfaction that has a positive impact on the revenue stream of RFID users. Naturally to create the mentioned value certain resources as well as capabilities and sometimes changes at organizational structures are required. As these requirements cause costs, it will be logical to invest in RFID if the impact of created value on revenue is greater than its impact on costs that arise to create the value. Only those organizations that consider this relation can benefit from RFID business models.

Technological developments contribute much to decrease especially the infrastructure costs of RFID technology. It is expected that this will increase the utilization of RFID in the near future. As a result of this companies will apply RFID technology to compete with their rivals. RFID adversaries declare that RFID threats data privacy. Despite this declaration it seems that the rapid deployment of RFID cannot be prevented, because both technical and regulative countermeasures are available to protect data privacy.

REFERENCES

1. F. Betz, 2002Strategic Business Models, Engineering Management Journal, 141March, 2002), 2127

2. J. Bohn, V. Coroamă, M. Langheinrich, F. Mattern, M. Rohs, 2004Living in a World of Smart Everyday Objects- Social, Economic, and Ethical Implications, Human and Ecological Risk Assessment, 105October, 2004), 7637851080-7039print / 1549-7680 online.

3. G. S. Erickson, E. P. Kelly, 2007Building Competitive Advantage With Radio Frequency Identification Tags, Competitiveness Review : An International Business Journal incorporating Journal of Global Competitiveness, 17137461059-5422

4. E. Filimon, 2008Anti-counterfeiting- prevention of counterfeit products with RFID, In: Business Aspects of the Internet of Things, Michahelles, F. (ed.), 1924ETH (Eidgenössische Technische Hochschule) Seminarreport, Zürich.

5. M. Hoffmann, D. Jerzynek, R. Weinand, 2005Fiktives Gutachten zum RFID-Einsatz im Einkaufszentrum Mitte im Auftrag des Berliner Senat, 17.01.2011, Available from http://ig.cs.tu-berlin.de/lehre/w2004/ir1/uebref/gutachten/HoffWeinJerz-Rfid-einsatzImEinkaufszentrumMitte-2005-02-09.pdf.

6. F. Kamoun, 2008Rethinking the Business Model with RFID, Communications of the Association for Information Systems (CAIS), 22Article 35, (June, 2008), 636658

7. T. Karygiannis, B. Eydt, G. Barber, L. Bunn, T. Phillips, 2007Guidelines for Securing Radio Frequency Identification (RFID) Systems- Recommendations of the National Institute of Standards and Technology (NIST), NIST Special Publication 80098Computer Security Division Information Technology Laboratory National Institute of Standards and Technology, Gaithersburg.

8. A. Kavas, 2007Radyo Frekans Tanımlama Sistemleri, Elektrik Mühendisliği Dergisi, Sayı 430, (Nisan, 2007), 7480

9. M. Kış, 2006RFID'nin Geleceği, 21.01.2011, Available from http://kodveus.blogspot.com/2006/05/rfidnin-gelecei.html.

10. E. Korkmaz, A. Üstündağ, M. Tanyaş, 2006Standards, Security & Privacy Issues about Radio Frequency Identification (RFID),

Proceedings of the 4th International Logistics and Supply Chain Congress, 9-75878-908-22930November and 1 December 2006.

11. LakeWest Group, LLC & MeadWestvaco Intelligent Systems2003RFID In Retail- The Future Is Now, 11.06.2008, Available from http://www.lakewest.com/PDFdocs/RFID%20 In%20Retail%20The%20Future%20Is%20Now_June%202003.pdf.

12. H. Lin, T. , L. , W. , S. , C. Chiang, L. , 2006Using RFID in Supply Chain Management for Customer Service, Proceedings of 2006 IEEE International Conference on Systems, Man and Cybernetics, 137713811-42440-099-6Taiwan, October 8-11, 2006.

13. A. Osterwalder, Y. Pigneur, C. L. Tucci, 2005Clarifying Business Models: Origins, Present, and Future of the Concept, Communications of the Association for Information Systems (CAIS), 15Tutorial, (May 2005).

14. F. F. Raafat, W. R. Sherrard, L. Meslis, J. Windt, 2007Case Study: Applications of RFID in Retail Business, 25.01.2011, Available from http://www.swdsi.org/swdsi07/2007_proceedings/papers/695.pdf.

15. Y. Shi, T. Manning, 2009Understanding Business Models and Business Model Risks, The Journal of Private Equity, 122Spring 2009), 4959

16. L. Schweizer, 2005Concept and Evolution of Business Models, Journal of General Management, 312Winter 2005), 3756

17. A. Üstündağ, 2008RFID ve Tedarik Zinciri, Sistem Yayıncılık, 978-9-75322-515-1stanbul

18. T. Weber, 2003The Future of Shopping, BBC News Article, 04.01.2011, Available from http://news.bbc.co.uk/2/hi/ business/ 3712261.stm.

CHAPTER 2

The Role of RFID Technology in Supply Chain Risk Management

May Tajima[1]

[1] The University of Western Ontario, Canada

1. INTRODUCTION

Supply chain risks come in a variety of forms: disruptions to material flows, product quality problems, information systems breakdowns, and economic instability (Chopra & Sodhi, 2004; Zsidisin et al., 2000). The recent literature in supply chain management recognizes the importance of managing such risks in the age of global supply chains. Various researchers have discussed firms' increasing exposure to risks and the resulting, potentially severe negative impact on the firms' financial performances (e.g., Hendricks & Singhal, 2005).

One such risk to the supply chain, disruption of supply flows, can occur suddenly due to a number of unpredictable events. Even more unpredictable, however, is the ripple effect caused by the disruption. For example, the September 11th terrorist attacks of 2001 in New York and Washington, D.C., originally disrupted many supply chains on the United States (U.S.) East Coast, one of which was the Ford Motor

Company's parts supply chain. The disruption eventually forced not one but five of Ford's assembly plants to cease production within a week of the incident (Zakaria, 2001). While Ford was experiencing parts shortages, Quanta Computer, a Taiwanese contract manufacturer for Dell and others, faced a pile-up of finished products when the U.S. airspace closed due to the attacks (Einhorn, 2001). One logistics service company in Europe estimated that the attacks cost the company £5 million (Parker, 2002). In this example, the ripple effects were extensive, affecting businesses in North America, Asia, and Europe. This high degree of impact clearly illustrates the importance of managing ripple effects as a part of supply chain risk management. In the first of two parts, this research shows that Radio Frequency Identification (RFID) technology, a relatively new development in supply chain management, holds great promise for managing supply disruptions and for containing their harmful ripple effects.

RFID a wireless technology that uses transmitted radio signals to tag an item in order to track and trace its movement without human intervention has superior capabilities over bar codes and promises many supply chain benefits, such as reductions in shrinkage, efficient handling of materials, increased product availability, and improved asset management (Angeles, 2005; Li & Visich, 2006; Taghaboni-Dutta & Velthouse, 2006). RFID has many applications in retail, healthcare, logistics, records management, and more, but so far its use in risk management has not been explored in the literature. To fill that gap, this research first addresses the following question:

Is RFID applicable in supply chain risk management; in particular, how is it useful for managing supply disruptions?

Based on RFID's technological capabilities, this research identifies three areas in which this technology could be utilized in the management of supply disruption risk: (i) monitoring for a disruption, (ii) responsiveness to the disruption, and (iii) the quality of decision-making involved in choosing corrective actions. Each of these three areas is discussed with a particular focus on how RFID could help to reduce the harmful ripple effects that are generated from supply disruptions. In order to provide support for these uses of RFID in risk management, this research presents case studies that originated from newspaper, magazine, and journal articles.

The discussion on RFID's risk management capabilities considers RFID as a source of advantages for firms that adopt the technology. However, the unprecedented level of supply chain visibility that is possible by the use of RFID can also be a source of risk. The literature has identified a number of concerns about this high degree of RFID-enabled visibility into supply chain activities. The concerns include consumer privacy invasion, corporate system security concerns, and industrial espionage (e.g., Juels, 2006; Shih et al., 2005). The second question in this research draws its motivation from the need to look at the other side of the same coin in order to gain a full understanding of RFID technology within the context of supply chain risk management:

What are the specific risks associated with RFID-enabled supply chain visibility, and how can these risks be mitigated?

The concerns associated with RFID's capability to provide supply chain visibility represent a timely and important research topic because similar concerns have been raised for other technologies that are capable of collecting, storing, and accessing huge amounts of data on individual items or people. For example, the Quit Facebook Day event in 2010 demonstrated Facebook members' concern for the privacy breach by the world's largest social networking website (CNN, 2010), which is capable of generating an unprecedented level of visibility into personal relationships. In the second part of this research, a review of the literature is conducted to identify specific risks associated with RFID's capability to provide supply chain visibility, and the research goes on to examine the existing mitigation approaches for dealing with RFID's visibility-related risks. Finally, the management implications are provided for the use of RFID in supply chain risk management based on both advantages and risks of its use.

The remainder of this chapter is organized as follows. Section 2 provides a review of the background literature. Section 3 presents the first part of this research, which focuses on RFID as a source of advantages in supply chain risk management, and Section 4 presents the second part, which focuses on RFID as a source of risks. Section 5 concludes the chapter with a summary of research contributions, limitations, and directions for future research.

2. BACKGROUND

This section provides background information for this research. Two areas of the literature are particularly relevant: Section 2.1 reviews the capabilities and applications of RFID technology, and Section 2.2 reviews those risk management elements that are associated with supply disruptions.

2.1. RFID capabilities

RFID is an automatic identification technology that identifies specific items and gathers data on them without human intervention or data entry (Wyld, 2006). Item identification occurs when a reader scans an RFID tag that is tuned to the same frequency as that of the reader. Fundamentally, RFID technology can be summarized by the following characteristics: (a) RFID is wireless, (b) it provides unique identification to an object, and (c) it traces and tracks objects (Kärkkäinen & Holmström, 2002). Each of these fundamental characteristics leads to an advantage over the existing bar code technology and allows RFID to possess three distinct capabilities: (i) advanced process automation, (ii) closed-loop tracking, and (iii) supply chain visibility (Tajima, 2007). These capabilities and their related applications are discussed in turn, below.

First, RFID's wireless characteristic eliminates the need for product positioning that is associated with bar-code scanning. This allows for the contents of mixed pallets to be identified simultaneously without undoing the packaging. Hence, compared to bar codes, RFID can support a higher degree of automated material inspection and handling (McFarlane & Sheffi, 2003). This process-automation capability provides many benefits in the management of warehouses and logistics by reducing material handling time and human errors in operations, such as receiving, inventory counting, data entry, put-away, routing for cross-docking, and custom clearance for cross-border shipments (Rutner et al., 2004; Zebra Technologies, 2004).

Second, RFID's ability to provide a unique identifier to an object comes from the fact that an RFID tag has a higher data capacity than does a bar code. This higher data capacity provides RFID with advanced record keeping and retrieval capability, through which RFID enables closed-loop tracking of individual items and assets, an action that is not possible with bar codes, which refer only to a class of products (Wyld, 2006). Recently, a wide range of applications has been identified for RFID's closed-loop tracking, including the tracking of medical devices within a hospital; paper documents within a law firm; gaming chips in casinos; media players for rental cars; and flower-growing operations from seeds to blooms (RFID Update, 2006c, 2007a, 2007b, 2007d, 2008b).

Third, RFID's ability to track and trace objects provides supply chain-wide, real-time visibility of individual items. When combined with other real-time locating technologies, such as Global Positioning Systems (GPS), RFID can be used to capture product information such as a detailed description of the product, its manufacture and expiration dates, the time of its departure and arrival at various facilities, and the address and telephone number of its manufacturer (EPCglobal, 2004). RFID-generated product information can provide an unprecedented level of visibility in the supply chain when shared among supply chain partners, a level of visibility that is simply not obtainable from bar codes. In the retail industry, where inaccuracy of inventory data is a major problem (Raman et al., 2001), one of the major applications of RFID is to improve inventory visibility. RFID can also increase the visibility into shipment data, which can in turn improve demand visibility (Lapide, 2004; McCrea, 2005). Automatic replenishment using "smart shelves" is another application in the retail industry and is considered valuable by, for example, German retailer METRO and Finnish apparel manufacturer NP Collection (RFID Update, 2007c, 2007e). For the pharmaceutical industry, the degree of supply chain visibility provided by RFID is considered critical for anti-counterfeiting measures and product recall management (Wicks et al., 2006; Wyld & Jones, 2007).

As shown above, RFID technology has applications in a wide range of industries and settings, but it has not yet found a place in the area of risk management.

2.2. Supply chain risk management

As mentioned in the Introduction, supply chain risks come in a variety of forms. To limit the scope of discussion, however, this research focuses solely on supply disruptions. In this research, supply disruptions are, as defined in Craighead et al. (2007), the disruptions of the normal flows of goods and materials within a supply chain that are caused by unplanned and unanticipated events. These disrupting events come in the various forms, such as natural disasters, labor disputes, wars, power failures, supplier contract breaches, and infectious diseases (Chopra & Sodhi, 2004; Haksöz & Kadam, 2009; Tang, 2006). For the purpose of this research, a ripple effect of a disruption is defined as any other supply disruptions that occur at different locations and/or at later dates due to the original disruption.

Typical risk management consists of four elements: (i) risk source/driver identification, (ii) risk consequence and likelihood assessment, (iii) risk mitigation and treatment, and (iv) risk monitoring. For risk source identification, Helferich (2002) indicated that supply disruptions could occur from interruptions in production facilities, supplier networks, transportation networks, communication infrastructure, and electricity and water services. Global sourcing is particularly vulnerable to supply disruptions because it generally involves greater distance, longer transit time, limited transportation mode, and complex security protocols for border crossings (Prater et al., 2001; Zsidisin, 2003). The just-in-time system is also susceptible to supply disruptions because it operates under fast-cycle procurement and lean inventory (Aichlmayr, 2001).

For risk assessment, Haksöz and Kadam (2009) studied ways to assess the supply disruption risk that results from supplier contract breaches. In their study, a tool to assess the financial impact of contract breaches was developed.

Risk mitigation focuses on ways to avoid, reduce, eliminate, buffer, or hedge against risk. A variety of operational strategies for mitigating supply disruptions have been examined in the literature. Chopra and Sodhi (2004) discussed having redundant suppliers, adding capacity, and increasing responsiveness as possible mitigation strategies. Sheffi (2001) proposed a multiple sourcing strategy that allocates the bulk of the procurement volume to inexpensive offshore suppliers but also gives a fraction of the business to local suppliers as insurance against supply disruption. Prater et al. (2001) identified a number of advantages in using local logistics operators, such as their knowledge of regional transportation routes and their familiarity with the border-crossing procedures. Babich et al. (2007) studied a hedging strategy based on the pricing and ordering policies of multiple suppliers.

Some authors studied inventory-related strategies for mitigating supply disruptions. For example, Sheffi (2001) discussed the emergency designation for safety stock in order to discourage its use for day-to-day fluctuations. Martha and Subbakrishna (2001) suggested increasing safety stock for critical items only, such as those coming from a single international source or those whose shortage quickly leads to plant shutdowns. When transfer or production of goods is not possible within a reasonable time frame, a marketing strategy may be used to steer customers toward substitutes. This strategy was exercised by Dell in response to the September 11th terrorist attacks (Rocks, 2001): Dell salespeople searched online to see which configurations of computers were available and then steered customers accordingly. Finally, Craighead et al. (2007) identified two key capabilities for mitigating supply disruptions: the capability to detect and disseminate information pertaining to the disruptive event, and the capability to respond quickly and effectively to the disruption.

The next section, which presents the first of two parts in this research, highlights RFID's usefulness in supply chain risk management by demonstrating that RFID can improve some of the risk-mitigation strategies mentioned above.

3. RFID AS A SOURCE OF ADVANTAGES

Is RFID applicable in supply chain risk management, and in particular, how is it useful for managing supply disruptions? This section addresses this research question by showing that an understanding of RFID's technological capabilities can lead to the discovery of RFID's risk management capabilities.

3.1. Methodology

First, it is shown that RFID's technological capabilities of closed-loop tracking, process automation, and supply chain visibility yield three specific risk management capabilities: increased monitoring capacity, increased response speed, and higher decision-making quality. Then, case studies are presented for all three risk management capabilities in order to provide support for their validity. As shown in Section 2.1, RFID application that is specific to the area of risk management has not yet been explored. Hence, any related case studies that could highlight the potential use of RFID in risk management were searched from newspaper, magazine, and journal articles. Below, RFID's three risk management capabilities are discussed in turn.

3.2. Monitoring capacity

Risk monitoring, as discussed in Section 2.2, is one of the typical elements in risk management, and it plays an important role in the management of unexpected supply disruptions. With an ability to monitor for and detect a disruption as it happens, corrective actions can begin sooner, the escalation of the disruption can be avoided, and the impact of the disruption, direct or indirect, can be reduced. Craighead et al. (2007) identified risk monitoring as one of the key capabilities needed for mitigating supply disruptions. It is shown below that closed-loop tracking, one of RFID's technological capabilities, can increase a firm's risk monitoring capacity.

As discussed in Section 2.1, the data capacity of RFID tags is higher than that of bar codes, and this higher data capacity allows for the closed-loop tracking of individual items and assets. RFID can be used to monitor not only cases and pallets but also individual raw materials, work-in-process inventories, and finished products. It can also monitor the use and condition of equipment and reusable assets. Therefore, with its closed-loop tracking capability, RFID can increase a firm's monitoring capacity by increasing the level of details that can be monitored.

The following case studies provide support for RFID's ability to increase a firm's monitoring capacity. At Nestlé, a large global food company, RFID was used to track the cleanliness of product trays (Bear, Stearns & Co. Inc., 2003). Such RFID-enabled tracking of reusable assets would extend Nestlé's capacity for detecting poor product quality to include the work-in-process items in addition to finished products. At the Wynn Hotel and Casino in Las Vegas, poker chips imprinted with RFID were used to monitor game play for possible cheating or gambling addiction (Wyld, 2008). In this case, RFID would increase the casino's capacity to detect problematic gaming behavior from the table/station level to the individual player.

RFID's closed-loop tracking capability can also increase a firm's risk monitoring capacity by providing the firm with an ability to monitor huge volumes of assets. RFID has already successfully managed a variety of assets with huge volumes. For example, a casino tracked 80,000 uniforms through the laundry process, and a beer company tracked three million beer kegs using RFID (Bear, Stearns & Co. Inc., 2003; Byrne, 2004). Byblos Amoreiras, a Portuguese book retailer, used RFID to track 150,000 books, periodicals, CDs, and other merchandise in its store (RFID Update, 2008a).

By increasing a firm's risk monitoring capacity, RFID can assist the firm with the identification of critical items. In Section 2.2, some risk mitigation strategies, such as an increase of safety stock and multiple sourcing, were discussed for critical items (Martha & Subbakrishna, 2001). With the use of RFID, firms could quickly identify critical items, such as those that run out first or those whose shortage causes a plant shutdown. Also, RFID's capability to monitor huge volumes of items can assist firms with the collection of historical data at the individual item level. These data would be useful in improving a firm's risk assessment in terms of estimating and updating the severity and likelihood of various supply disruptions.

3.3. Response speed

The previous section focused on RFID's ability to detect a disruption at the level of individual items and assets. Once detected, a firm's ability to respond quickly to the disruption becomes important in the containment of the ripple effects. Responsiveness has been identified as one of the key capabilities for managing supply disruptions in the literature (Chopra & Sodhi, 2004; Craighead et al., 2007). Another of RFID's technological capabilities, process automation, can increase a firm's response speed.

As discussed in Section 2.1, when compared to bar code technology, RFID's wireless characteristic allows for a higher degree of automation in the processes, such as material inspection and handling (McFarlane & Sheffi, 2003). For LCWaikiki, one of the largest apparel retailers in Turkey, the replacement of bar codes with RFID technology has resulted in the merchandise transfer from the back room to the shop floor being performed 70% faster and the merchandise receiving being performed 60% faster (RFID Update, 2008d). For Bloomingdale's, a large U.S. department store, an RFID pilot study resulted in a 96% reduction of cycle counting time for the store's inventories (RFID Update, 2009). For American Apparel, a large U.S. clothing manufacturer and retailer, with the use of RFID, the time required for store-level inventory count dropped from 120 work-hours to 15 work-hours (Avery Dennison Corporation, 2010). These case studies support RFID's ability to speed up some of the common responses to a supply disruption, such as recounting inventories, adjusting shipment data, and sending invoice reconciliations.

In reality, the response to a supply disruption cannot begin until key personnel within a firm are notified of the disruption. Once notified, these individuals can then authorize the start of corrective actions. In this leg of the process, RFID's process automation capability can increase a firm's response speed by facilitating the real-time alert for notifying key personnel in the event of a supply disruption. Throttleman, a Portuguese fashion retailer, has set up a real-time alert system using RFID in its distribution center (RFID Journal, 2007). Upon arrival at the distribution center, the contents of a box are automatically identified using RFID without opening the box. The captured contents are then compared to the items listed in an advance shipping notice that has been electronically sent by the garment manufacturer. If the received contents do not match with the advance shipping notice, then an alarm goes off for the center's personnel to physically deal with the discrepancy. In another instance, a real-time alert system has been implemented at several U.S. hospitals to notify staff immediately when a piece of equipment becomes misplaced (Emrich, 2008). Also, at Lincoln University, a Pennsylvania liberal arts college, valuable audio-visual equipment was tracked using RFID, and an alert notified the IT department as soon as a piece of equipment left its predetermined zone (RFID News, 2008). These case studies support RFID's ability to increase a firm's response speed by setting up a real-time personnel alert system.

3.4. Decision-making quality

Upon notification of a supply disruption, key personnel need to assess the extent of the disruption and decide on the appropriate risk mitigation strategies before corrective actions can actually begin. The quality of these strategic decisions can have a significant impact on the outcome of the corrective actions. For example, Hurricane Mitch in 1998 caused a supply interruption for two major banana producers in Central America, Dole and Chiquita (Martha & Subbakrishna, 2001). Dole's business suffered from this supply disruption, with the subsequent ripple effect lasting longer than a year. Chiquita, on the other hand, had a significantly different outcome: it was able to arrange alternative supply sources, and its revenue actually grew during the last quarter of 1998.

In general, management decisions are often made based on incomplete or old data (Lin et al., 2006). Therefore, an overall increase in information accessibility has significant potential to improve the quality of management decisions, including the ones that must be made in response to supply disruptions. As discussed in Section 2.1, compared to bar codes, RFID promises an unprecedented visibility into supply chain operations. Supply chain-wide visibility provides information such as inventory levels, shipment data, locations of stockpiles, and alternative suppliers throughout the extended enterprise. Such information is critical in providing a firm with the ability to redirect its inventories within its supply chain, or to steer customers toward substitute products based on informed decision-making, rather than based on incomplete or untimely data. Hence, the third of RFID's technological capabilities, supply chain-wide visibility, can improve the quality of a firm's decision-making in the selection of risk mitigation strategies by increasing the completeness and timeliness of information available for the decision-makers.

Several case studies provide support for RFID's ability to improve the quality of a firm's risk mitigation decisions. In the retail industry, electronic article surveillance (EAS) devices provide retailers with the knowledge of timing when something is stolen from a store, but it cannot reveal which item has been taken (RFID Journal, 2009). RFID, on the other hand, can provide the retailers with more complete information. At Sony Europe, a combination of RFID, EAS, and a video surveillance system was implemented in its largest European distribution warehouse, located in the Netherlands (RFID Journal, 2009). The system was designed to deter employee or professional theft by giving Sony as much information as possible on each theft: which item is stolen, when it is stolen, and who may be doing the stealing. In another case, the additional data obtained through RFID allowed one retailer to successfully link multiple thefts over a period of time to a single person (Arnstein, 2010). Moreover, RFID-generated data can support a targeted and cost-effective security strategy that provides different security levels for different products within the same store, such as a silent alarm for expensive items and an audible alarm for inexpensive ones (Arnstein, 2010). These case studies support RFID's ability to improve the quality of a firm's risk mitigation decisions by increasing the completeness of the information available for the decision-makers.

As discussed in Section 2.1, when combined with other real-time locating technologies, such as GPS, RFID is capable of capturing product information within a supply chain on a real-time basis. Hence, in addition to the increase in completeness of information, RFID-enabled supply chain visibility can increase the timeliness of information available for the decision-makers in a firm. For example, through the use of RFID, Dole Food Company, the world's largest producer and marketer of fresh fruits and vegetables, was able to initiate a voluntary, pre-emptive recall of packaged salads that were suspected of E. coli bacteria contamination *before* any consumers were reported ill (Uldrich, 2007). When the recall announcement was made, Dole also knew that a total of 5,058 bags of salad were most likely to have been exposed to the bacteria, of which 528 bags were distributed in Canada and 4,530 bags were distributed within eight U.S. states. The value of RFID in providing timely information was also discussed in a simulation study conducted by Kim et al. (2010). Their study showed that an RFID-based, vehicle-tracking system could significantly decrease the overall transfer time of finished vehicles from an automobile assembly plant to its shipment yard by providing the real-time availability of parking spots. The yard operators were then able to use real-time information to make their decisions more efficiently and effectively. Without RFID, the status of parking availability could be updated only periodically through a manual reporting process, and therefore, the yard operators had to make their decisions based on untimely data.

All the related case studies presented above support RFID's ability to improve a firm's risk management capabilities in terms of its monitoring capacity, response speed, and decision-making quality. As a result, this section clearly demonstrates RFID's applicability in supply chain risk management and its usefulness in managing supply disruptions.

4. RFID AS A SOURCE OF RISKS

The previous section focused on RFID as a source of advantages for firms that adopt the technology and use it for supply chain risk management. However, the use of RFID can also be a source of various risks in and of itself. Within the literature, numerous articles have discussed RFID-enabled supply chain visibility as a source of security and privacy risks. The concerns surrounding these risks are a timely and important research topic, from both industry and society perspectives.

From a society perspective, a general feeling of anxiety exists toward technology-enabled information visibility. As mentioned in the Introduction, the Quit Facebook Day event in 2010 stands out as a high-profile example of people's concerns for the breach of users' personal information. Another example concerns Google's Street View, which provides panoramic views of streets all over the world, as captured by a fleet of vehicles that are equipped with high-tech cameras and scanners. The main concern for this technology stems from the fact that these panoramic images are publicly accessible from the Google website and may contain personally identifiable details, such as people's faces, belongings, and cars on the driveways with visible license plate numbers, all matched to easily identifiable street addresses (Bradley, 2010). Digital medical records represent another high-profile example of the public's concerns for technology-enabled information visibility. On one hand, electronically accessible medical records offer many benefits, including the reduction of duplicate diagnostic testings and medical errors. On the other hand, medical records consist of highly personal information, from prescription records to family-health histories, and may even include DNA information in the future (Fox News, 2010). A concern exists among physicians in terms of how to protect such private, sensitive, and massive data from potential hacking. As can be seen from these examples, the concerns for technology-enabled information visibility are not unique to RFID, and at present, no absolute solutions or countermeasures exist to deal with these concerns.

From an industry perspective, the risks associated with RFID-enabled supply chain visibility constitute a timely and important research topic for RFID vendors, potential users, and corporate and public policy-makers mainly because RFID is still a developing technology (RFID Update, 2008c), whose industry adoption may easily be hindered by any risk related to its use. With other established supply chain technologies, such as bar codes, electronic data interchange (EDI), and enterprise resource planning (ERP), information sharing and the resulting visibility have not posed any serious issues since the scope of visibility has rarely been extended to involve individual items or consumers. Consequently, within the context of supply chain technologies, RFID has no obvious precedence to follow regarding how to deal with the risks related to information visibility, and this makes an understanding of these risks critical for RFID technology's future growth.

What, then, are the specific risks associated with RFID-enabled supply chain visibility, and how can these risks be mitigated? The remainder of this chapter focuses on addressing this research question.

4.1. Methodology

A review of published literature is provided on possible risks associated with RFID-enabled supply chain visibility. Two databases, ProQuest and Scholars Portal, were used to search relevant articles. The chosen search terms utilized various combinations of: RFID, risk, security, and privacy. The search dates were restricted to the years between 2003 and 2010. Due to insufficient resources for translation, the review was also restricted to English-language articles only. The search produced over 100 articles, covering more than 50 different journals from a variety of disciplines, such as business, engineering, information systems, economics, law, electronic commerce, marketing, production, and healthcare. Therefore, although the search was not exhaustive, the search range was considered sufficiently comprehensive in terms of the variety of articles, and further searches from other databases were deemed unnecessary.

Based on the articles found in the search described above, Section 4.2 provides the first result: an overview of two main categories of RFID's supply chain visibility risks, which are security risks and privacy risks. Section 4.3 then provides the second result: a classification of the existing mitigation approaches for dealing with RFID's supply chain visibility risks. Finally, Section 4.4 discusses the management implications of the use of RFID in supply chain risk management based on its advantages as well as risks.

4.2. Supply chain visibility risks

In the RFID literature, a variety of risks associated with the use of RFID have been discussed. For example, a risk to patients' health that might result from the altering of the chemical composition of a medication was discussed in the context of using RFID for pharmaceutical products (Symbol Technologies, 2006). A risk to the environment was discussed in relation to the disposal of non-biodegradable RFID tags (Li & Visich, 2006). A risk to the corporate information system was also discussed in terms of the vulnerability of RFID to computer viruses (RFID Update, 2006b). This section, however, focuses on providing an overview of the risks that are specifically associated with RFID-enabled supply chain visibility: security risks and privacy risks.

Security risks. Security risks in the RFID literature are often discussed as attacks against organizations by their competitors, opponents, or criminals. Various types of attacks are possible with the use of RFID. One type is referred to as "data eavesdropping," which is the interception of communications between RFID tags and readers. Through data eavesdropping, a military security breach may occur if enemy forces detect troop locations and monitor their movements by tracking RFID tags within a military supply chain (Juels, 2006; Zuo, 2010). Corporate espionage is also possible through data eavesdropping. For example, by tracking RFID tags through the retail supply chain, competitors may spy on one retailer's sensitive business data, such as sales trends, pricing trends, stock selections, and stock turnover rates (Juels, 2006; Li & Visich, 2006; Shih et al., 2005). A seller organization may also attempt to gain visibility into the downstream of the supply chain by monitoring RFID tags on the sold items after the seller no longer has physical access to the items (Kapoor et al., 2009).

Another type of attack against organizations is referred to as "data corruption," which erases or modifies RFID tag contents. If the tag contents include price information, then, through data corruption, hackers could lower the price of expensive retail items, and then use an RFID-enabled self-checkout counter to avoid detection by store employees (Li & Visich, 2006). Spoofing, another type of attack, involves the retrieval of confidential information by impersonating authentic readers (Shih et al., 2005). Spoofing can lead to, for example, counterfeiting of retail products by falsely authenticating fake products using stolen authentication information. Finally, denial of service is a type of attack that renders RFID tags temporarily or permanently incapacitated (Zuo, 2010). Denial of service can cause a loss of business data and operational disruptions to an organization.

Privacy risks. While security risks typically affect organizations and result in financial losses, privacy risks affect individuals and result in ethical issues. The literature discusses three main issues that are specifically related to RFID's ability to provide supply chain visibility that includes end-consumer information.

The first issue relates to the collection of personal data without an individual's knowledge or consent. This concern stems from the fact that the size of RFID tags can be as small as grains of sand, making it possible to inconspicuously attach the tags on products. Also, the scanning of RFID tags is a wireless process that cannot be detected by human eyes or ears. Hence, a retailer is technically able to conduct market research, for example, by tracking RFID tags on pre-sale items inside the store without the knowledge or consent of the consumers (Jones et al., 2004).

The second ethical issue relates to the infringement on individual anonymity. In the context of supply chain management, RFID tags are traditionally associated with product information but not with consumer information. However, since an RFID tag is capable of providing a unique identifier to a product, any association between the product and an individual can in turn become the unique identifier of the individual. For example, a female customer with a previously purchased item carried in a purse can be identified as a returning customer if the tag on the item is read upon her return to the store. Even if the retailer does not possess full information on her identity, the anonymity of this customer can still be infringed upon since it is possible to build a personal profile based on information such as the frequency of the store visit, the time and day of the visit, and the history of other purchases made by this customer (Wasieleski & Gal-Or, 2008). Such consumer profiles could then be exploited for price differentiation strategies or could be sold to third parties (Jones et al., 2004; Peslak, 2005). Moreover, if the item in question was, for example, a prescription drug bottle being carried by an individual, then the product information itself could represent a piece of sensitive personal data.

The third ethical issue, the surveillance of individuals, also stems from the association made between a product and an individual by the use of RFID. By tracking RFID tags on products that are owned by individuals, people may be tracked in the stores, on the streets, and even in people's homes (Jones et al., 2004; Rutner et al., 2004). This issue is often discussed in relation to the idea of "Big Brother," where the authority monitors civilians' every move.

All the ethical issues discussed above have been fueling an opposition to RFID from various consumer advocacy groups, such as Consumers Against Supermarket Privacy Invasion and Numbering (CASPIAN), the American Civil Liberties Union, the Privacy Rights Clearinghouse (PRC), all of which are generally against the use of RFID (Barut et al., 2006; Jones et al., 2004).

4.3. Risk mitigation approaches

The actual occurrence of security and privacy risks is still not common due to the limited and fragmented use of RFID, and the literature discusses these issues as potential risks of RFID. However, as mentioned previously, any risk related to RFID's use may negatively affect the growth of this technology. How, then, can these risks be mitigated? This section presents a classification of the existing mitigation approaches for dealing with RFID's supply chain visibility risks. An examination of the RFID literature revealed that four general treatments currently exist: technology-based countermeasures, business policies, consumer education, and legal measures. Each of these approaches is discussed below.

Technology-based countermeasures. In order to reduce the likelihood of the occurrence of security and privacy risks, some technology-based countermeasures have been proposed in the literature. One group of countermeasures is designed to protect RFID tags from unauthorized scanning, including: tag killing, to make tags permanently inoperative when the tags receive a "kill" command from a reader; tag sleeping, to make tags temporary inactive unless "woken" by authorized users; tag relabeling, to give tags different identifiers periodically; tag encryption, to use cryptography to encrypt tag data or identifiers; and hash locks, to make tags respond to data queries with only limited information when "locked" (e.g., Juels, 2006; Shih et al., 2005; Zuo, 2010). By protecting the tags, these countermeasures are intended to prevent security attacks, such as data corruption, spoofing, and denial of service; they can also prevent the surveillance of individuals.

Another group of technology-based countermeasures is designed to protect communications between RFID tags and readers. Most of these countermeasures are based on developing protocols for the search and authentication procedures that occur between the tags and readers (Zuo, 2010). Another approach to protect the tag-reader communication is to limit the tag-reading area, which can be accomplished by tag clipping, a process that shortens the antenna in a tag to reduce its read range (Kapoor et al., 2009), or by shielding the tag-reading area with metal screens to prevent the unauthorized scanning from outside (Swartz, 2007). The protection of communications between tags and readers would certainly be useful for the prevention of attacks that involve data eavesdropping and spoofing.

Another group of technology-based countermeasures focuses on informing individuals about unauthorized scanning. For example, a watchdog tag is supposed to be carried by an individual to monitor for any unsolicited scannings against the individual (Juels, 2006). A read-write tag, also carried by an individual, keeps a log of unauthorized scannings (Li & Visich, 2006), and a blocking tag on an individual is supposed to block any unsolicited scannings (Juels, 2006). By alerting individuals about the practice of unauthorized scanning, these countermeasures give consumers an opportunity to, for example, look for a suspicious reader or walk away from the area in question, therefore, helping to alleviate the risks, such as data eavesdropping, collection of personal data without the individuals' knowledge or consent, and the surveillance of individuals.

Business policies. The use of business policies offers another general approach for dealing with RFID's supply chain visibility risks. While technology-based countermeasures address both security and privacy risks, the business policies discussed in the literature focus on dealing with privacy risks. Common policies on the use of RFID include making RFID tags clearly visible to consumers, making tags easily removable by placing them on the product packaging or on price tags, and disabling tags at the point of product purchase (Jones et al., 2004; Li & Visich,

2006; Taghaboni-Dutta & Velthouse, 2006). As an example, H.D. Smith, a major pharmaceutical product distributor, officially requested that its customers (i.e., pharmacies and hospitals) remove RFID tags upon receipt of shipments from H.D. Smith (Downey, 2006). Business policies may also include statements about a firm's data-collection practices. For example, Pfizer, a large pharmaceutical manufacturer, made a public statement that it would not collect any patient information using RFID (RFID Update, 2006a). Such policy statements are used by firms mainly to reassure consumers that the subject firms intend to use RFID responsibly, thereby minimizing the privacy risks associated with its use.

Consumer education. Another general approach for dealing with RFID's supply chain visibility risks comes in the form of consumer education. As with the business policy approach, consumer education focuses on dealing with privacy risks. In 2004, a survey of 1,000 North American consumers showed that only one in four knew what RFID was (Jones et al., 2004). A study on the consumer attitude toward RFID revealed that, the less consumers were educated about RFID, the more hesitant they were about the use of RFID in businesses (Razzouk et al., 2008). In light of a general lack of consumer understanding regarding RFID, an industry group, the Association for Automatic Identification and Mobility (AIM), has recognized the need to convey accurate information about the technology to the community (Peslak, 2005). Information on RFID's technical capabilities, as well as its limitations and comparisons of RFID to other wireless technologies, could educate consumers on the likelihood of privacy risks. Moreover, consumers could be further educated on the likelihood of privacy risks through access to information on whether an RFID tag is embedded in a product, when the tag is read, and whether the tag is removed or deactivated upon purchase (Pottie, 2004).

Legal measures. The final approach that is discussed in the literature as a way to deal with RFID's supply chain visibility risks is the legal approach. This approach mainly focuses on the protection of information privacy, which is the right of an individual to retain control over the collection and use of personally identifiable information (Kelly & Erickson, 2005). In 2004, California passed a bill prohibiting the use of RFID to collect, store, use, or share personal information unless certain legal conditions were met (Taghaboni-Dutta & Velthouse, 2006). A proposal to extend the Fair Information Practices, originally promoted by the Federal Trade Commission in the U.S. to protect online privacy, has also been put forth for the use of RFID (Peslak, 2005). Fair Information Practices include business practices such as notifying consumers of the collection of personal information; giving consumers options concerning how information is used; giving consumers access to the collected information; providing security over the collected data; and providing penalties for non-compliance. The European Union does not have RFID-specific regulations. However, its existing regulations — the Data Protection Directive of 1995, the Electronic Commerce Directive of 2000, and the Privacy and Electronic Communications Directive of 2002 — do apply to the personal data collected by the use of RFID (Slettemeås, 2009). By establishing and enforcing the laws on RFID-generated data, the legal approach is intended to deter the occurrence of security and privacy risks and to provide individuals with a means of recourse in the case of a privacy breach.

4.4. Management implications

The first part of this two-part research demonstrated that RFID could be a source of tremendous advantage for firms that adopt the technology and use it for managing supply disruptions. The second part of this research, however, showed that RFID could also be a source of security and privacy risks, and attacks on the RFID system, such as data corruption and denial of service, may actually cause supply disruptions through a loss of business data or the disruptions to internal operations. Hence, from a management perspective, the use of RFID in supply chain risk management requires careful consideration of the risks in and of RFID itself. The security attacks in general may be alleviated by the use of technology-based countermeasures. However, the management must be aware that various shortcomings have been documented for these countermeasures. For example, tag killing can eliminate the security risks, but it also eliminates many post-sale benefits of having RFID tags on products, such as efficient warranty processing, easy handling of returns, and goods authentication (Juels, 2006). The use of cryptography may be computationally infeasible when RFID tags are employed on a mass scale (Zuo, 2010), and a security protocol is secure only until its loopholes are discovered (Kapoor et al., 2009).

Privacy risks associated with the use of RFID may not cause supply disruptions, but they may negatively influence the consumer attitude towards RFID (e.g., Slettemeås, 2009), thereby hindering the growth of RFID adoption. Hence, the management needs to address the privacy risks in general, but it must be aware that the mitigation of the privacy risks is a complex subject that requires a multi-faceted solution, for none of the existing mitigation approaches provides an all-encompassing solution on its own. Technology-based countermeasures, as mentioned above, come with various shortcomings. Business policies and consumer education do not actually prevent the incidence of unauthorized scanning. With the legal approach, the burden is placed on the plaintiff to prove that a privacy breach has taken place and it resulted in a high degree of shame, humiliation, mental illness, and so on (Willey, 2007).

A further examination of the privacy risks reveals that, when a firm considers the use of RFID in supply chain risk management, the collection of personal data without an individual's knowledge or consent and the surveillance of individuals may not be the first risks that the firm needs to address since the data utilized in the context of risk management are mostly inventory, shipment, equipment, asset, and supplier data, but not consumer data. On the other hand, since a unique product identifier given by an RFID tag can turn into a unique personal identifier as discussed previously, the infringement on individual anonymity should be addressed whenever a firm uses item-level information. In terms of risk mitigation, all of the current approaches focus on when or how to stop the collection of personal data, but none of them effectively address the infringement on individual anonymity since they do not focus on what to do with the data that are already collected.

Based on this research, two suggestions are made for mitigating the infringement on individual anonymity when firms consider the use of RFID for supply chain risk management. First, the firms can utilize the consumer education approach to clearly communicate specific benefits for consumers resulting from the better management of supply disruptions. Based on the three risk management capabilities of RFID discussed previously, the consumers can expect benefits such as fewer and shorter business disruptions experienced by the consumers and increased public safety in certain cases (e.g., food recalls). Second, in addition to the business policies on whether or not certain data will be collected, the firms should consider adding policies on how they intend to utilize the collected data in order to come across as the responsible users of RFID in the eyes of consumers. For example, a firm may state that it will collect item-level product data via RFID for the purpose of detecting and mitigating supply disruptions.

5. CONCLUSION

The first part of this research demonstrated that RFID's three risk management capabilities — monitoring capacity, response speed, and decision-making quality — were applicable and useful in the management of supply disruptions. The second part of this research showed that the security and privacy risks were associated with RFID-enabled supply chain visibility, and that four general mitigation approaches exist at present: technology-based countermeasures, business policies, consumer education, and legal measures. Together, the two parts of this research provided a comprehensive understanding of the use of RFID in the context of supply chain risk management.

The main limitation of this research is that some practical issues related to the use of RFID in a real-life setting were not included in the discussion. One such issue is that a firm's corporate information system may not be capable of supporting the increased monitoring capacity that is promised by the use of RFID. A general increase of data processing needs has been discussed as one of the challenges associated with RFID implementation (Angeles, 2005). Another issue arises with the design of a real-time alerting system. The alerting of top personnel should ideally be reserved for severe supply disruptions only. This implies that the alerting system must recognize different levels of the disruptions in order to alert different levels of personnel. However, a firm may not have sufficient data on the actual supply disruptions with varying degrees. Also, as RFID improves the completeness and timeliness of information available for the key decision-makers, information overload may become an issue in general. Having abundant information may lead to the generation of many options for the decision-makers to assess, and consequently, it may slow the response to a supply disruption. Future research must address such practical issues in order to make RFID-based risk management a reality.

In conclusion, this research provided valuable insight into a novel application of RFID technology in the area of supply chain risk management. This insight was built from the balanced understanding of RFID as a source of advantages as well as a source of risks.

ACKNOWLEDGEMENTS

The author wishes to acknowledge Lelanya Perryman for her assistance in this research as a research assistant, whose work was funded by the Start-Up Research Grant at The University of Western Ontario.

REFERENCES

1. M. Aichlmayr, 2001The future of JIT time will tell. Transportation and Distribution, 42121822

2. R. Angeles, 2005RFID technologies: supply-chain applications and implementation issues. Information Systems Management, 2215165

3. L. Arnstein, 2010How to serve the multi-channel consumer. Material Handling Management, December 1

4. Avery Dennison Corporation.2010American Apparel RFID case study: the challenges. Date of access, March 21, 2011, Available from: www.ibmd.averydennison.com/solutions/american-apparel-rfid-challenges.asp

5. V. Babich, A. Burnetas, P. H. Ritchken, 2007Competition and diversification effects in supply chains with supplier default risk. Manufacturing and Service Operations Management, 92123146

6. M. Barut, R. Brown, N. Freund, J. May, E. Reinhart, 2006RFID and corporate responsibility: hidden costs in RFID implementation. Business and Society Review, 1113287303

7. Stearns. Bear, Inc. Co, 2003Supply-chain technology: track(ing) to the future. Equity research report, Date of access, September 10, 2006, Available from: www.bearstearns.com/bscportal/pdfs/research/supplychain/technology_rfid.pdf

8. T. Bradley, 2010Google Street View raises privacy concerns again. In: PCWorld, Date of access, February 18, 2011, Available from: www.pcworld.com/printable/article/id,190279/printable.html

9. P. M. Byrne, 2004RFID: not just for Wal-Mart anymore. In: Logistics Management, September 1, Date of access, September 10, 2006, Available from: www.logisticsmgmt.com/archive

10. S. Chopra, M. Sodhi, 2004Managing risk to avoid supply-chain breakdown. MIT Sloan Management Review, 4615361

11. CNN.2010Some quitting Facebook as privacy concerns escalate, In: CNN Tech News, Date of access, February 18, 2011, Available from: http://articles.cnn.com/2010-05-13

12. C. W. Craighead, J. Blackhurst, M. J. Rungtusanatham, R. B. Handfield, 2007The severity of supply chain disruptions: design characteristics and mitigation capabilities. Decision Sciences, 381131156

13. L. Downey, 2006An Interview with RFID Trailblazer H.D. Smith. RFID Update, August 21, Date of access, March 21, 2011, Available from: www.rfidupdate.com/articles/home.php

14. B. Einhorn, 2001Laptop king; in a year that's decimated high tech, Taiwan's unstoppable Quanta is posting double-digit sales growth. Business Week, 375648

15. A. B. Emrich, 2008Wireless RFID product tracks hospital assets. Grand Rapids Business Journal, 265014

16. EPCglobal.2004The EPCglobal Network and the Global Data Synchronization Network (GDSN). Date of access, September 10, 2006, Available from: www.epcglobalinc.org/news/position_papers.html

17. Fox News.2010WikiLeaks breach raises concern about privacy of electronic medical records. In: Fox News, Date of access, February 18, 2011, Available from: www.foxnews.com/politics/2010/12/07

18. Ç. Haksöz, A. Kadam, 2009Supply portfolio risk. The Journal of Operational Risk, 415977

19. O. K. Helferich, 2002Securing the supply chain against disaster. Distribution Business Management Journal, 231014

20. K. B. Hendricks, V. R. Singhal, 2005An empirical analysis of the effect of supply chain disruptions on long-run stock price performance an equity risk of the firm. Production and Operations Management, 1413552

21. P. Jones, C. Clarke-Hill, D. Hillier, P. Shears, D. Comfort, 2004Radio frequency identification in retailing and privacy and public policy issues. Management Research News, 2784656

22. A. Juels, 2006RFID security and privacy: a research survey. IEEE Journal of Selected Areas in Communications, 242381394

23. G. Kapoor, W. Zhou, S. Piramuthu, 2009Challenges associated with RFID tag implementations in supply chains. European Journal of Information Systems, 186526533

24. M. Kärkkäinen, J. Holmström, 2002Wireless product identification: enabler for handling efficiency, customization and information sharing. Supply Chain Management: An International Journal, 74242252

25. E. P. Kelly, G. S. Erickson, 2005RFID tags: commercial applications v. privacy rights. Industrial Management and Data Systems, 1056703713

26. J. Kim, C. S. Ok, S. Kumara, S. T. Yee, 2010A market-based approach for dynamic vehicle deployment planning using radio frequency identification (RFID) information. International Journal of Production Economics, 1281235247

27. L. Lapide, 2004RFID: what's in it for the forecaster? Journal of Business Forecasting Methods and Systems, 2321619

28. S. Li, J. K. Visich, 2006Radio frequency identification: supply chain impact and implementation challenges. International Journal of Integrated Supply Management, 24407424

29. D. Lin, R. Barton, H. Bi, M. Freimer, 2006Challenges in RFID enabled supply chain management. Quality Progress, 39112328

30. J. Martha, S. Subbakrishna, 2001When just-in-time becomes just-in-case. Wall Street Journal, October 22, A18

31. B. Mc Crea, 2005Reliable Foods lives up to its name. In: Logistics Management, July 1, Date of access, September 10, 2006, Available from: www.logisticsmgmt.com/archive

32. D. Mc Farlane, Y. Sheffi, 2003The impact of automatic identification on supply chain operations. International Journal of Logistics Management, 141117

33. R. Parker, 2002Just-in-time freight plans survive 11 September crisis. Supply Management, 758

34. A. R. Peslak, 2005An ethical exploration of privacy and radio frequency identification. Journal of Business Ethics, 594327345

35. G. J. Pottie, 2004Privacy in the global e-village. Communications of the ACM, 4722123

36. E. Prater, M. Biehl, M. A. Smith, 2001International supply chain agility: tradeoffs between flexibility and uncertainty. International Journal of Operations and Production Management, 215823839

37. A. Raman, N. De Horatius, T. Zeynep, 2001Execution: the missing link in retail operations. California Management Review, 433136152

38. N. Razzouk, V. Seitz, M. Nicolaou, 2008Consumer concerns regarding RFID privacy: an empirical study. Journal of Global Business Technology, 416978

39. RFID Journal.2007Throttleman adopts item-level tagging. In: RFID Journal, Date of access, January 19, 2011, Available from: www.rfidjournal.com/article/print/3580

40. RFID Journal.2009Retail theft is up- can RFID help? In: RFID Journal, Date of access, January 19, 2011, Available from: www.rfidjournal.com/blog/entry/5372

41. RFID News.2008Schools using RFID for security enhancement. In: RFID News, Date of access, January 19, 2011, Available from: www.rfidnews.org/2008/09/09/schools-using-rfid-for-security-enhancement

42. RFID Update.2006aPfizer Shipping RFID-Tagged Viagra. In: RFID Update, January 9, Date of access, March 21, 2011, Available from:www.rfidupdate.comarticles/home.php.

43. RFID Update.2006bStudy: RFID vulnerable to viruses and worms. In: RFID Update, March 15, Date of access, March 21, 2011, Available from:www.rfidupdate.com/articles/home.php

44. RFID Update.2006cDutch casino operator bets on RFID chips. In: RFID Update, December 8, Date of access, March 21, 2011, Available from:www.rfidupdate.com/articles/home.php

45. RFID Update.2007aRFID blooms for Dutch flower tracking. In: RFID Update, July 18, Date of access, March 21, 2011, Available from:www.rfidupdate.com/articles/home.php

46. RFID Update.2007bHospital manages thousands of patient files with RFID. In: RFID Update, August 1, Date of access, March 21, 2011, Available from:www.rfidupdate.com/articles/home.php

47. RFID Update.2007cWhy METRO's item-level RFID deployment matters. In: RFID Update, September 24, Date of access, September 24, 2007, Available from: www.rfidupdate.com/articles/home.php

48. RFID Update.2007dRFID file tracking is heating up. In: RFID Update, November 29, Date of access, March 21, 2011, Available from: www.rfidupdate.com/articles/home.php

49. RFID Update.2007eFinnish retailer gets quick ROI on item-level RFID. In: RFID Update, November 16, Date of access, March 21, 2011, Available from:www.rfidupdate.com/articles/home.php

50. RFID Update.2008aWorld's largest item-level RFID application launches. In: RFID Update, Date of access, January 19, 2011, Available from:www.rfidupdate.com/articles/home.php

51. RFID Update.2008bRFID helps kids take Dora the Explorer on road trips. In: RFID Update, March 10, Date of access, March 21, 2011, Available from: www.rfidupdate.com/articles/home.php

52. RFID Update.2008cRFID market on target to reach $5.3B in 2008. In: RFID Update, June 12, Date of access, June 12, 2008, Available from: www.rfidupdate.com/articles/home.php

53. RFID Update.2008dTurkish retailer plans big item-level RFID expansion. In: RFID Update, Date of access, August 11, 2008, Available from:www.rfidupdate.com/articles/home.php

54. RFID Update.2009Bloomingdale's tracks strong results from RFID pilot. In: RFID Update, Date of access, January 19, 2011, Available from: www.rfidupdate.com/articles/home.php

55. D. Rocks, 2001The Net as a lifeline; a tough economic environment makes the Web even more important for companies attempting to cut costs, generate new revenues, and better serve customers. Business Week, 3755EB16

56. S. Rutner, M. A. Waller, J. T. Mentzer, 2004A practical look at RFID. In: Supply Chain Management Review, January/February, Date of access, September 10, 2006, Available from: www.manufacturing.net/scm/

57. Y. Sheffi, 2001Supply chain management under the threat of international terrorism. The International Journal of Logistics Management, 122111

58. D. H. Shih, C. Y. Lin, B. Lin, 2005RFID tags: privacy and security aspects. International Journal of Mobile Communications, 33214230

59. D. Slettemeås, 2009RFID-the "next step" in consumer-product relations or Orwellian nightmare?: challenges for research and policy. Journal of Consumer Policy, 323219244

60. N. Swartz, 2007NIST issues RFID guidelines. Information Management Journal, 4148

61. Symbol Technologies2006A prescription for RFID success in the pharmaceutical industry. Date of access, March 21, 2011, Available from: www.symbol.com/assets/files/rfid_uhf.pdf

62. F. Taghaboni-Dutta, B. Velthouse, 2006RFID technology is revolutionary: who should be involved in this game of tag? Academy of Management Perspectives, 2046578

63. M. Tajima, 2007Strategic value of RFID in supply chain management. Journal of Purchasing and Supply Management, 134261273

64. C. S. Tang, 2006Perspectives in supply chain risk management. International Journal of Production Economics, 1032451488

65. J. Uldrich, 2007Dole: let us thank RFID technology. In: The Motley Fool, Date of access, February 17, 2011, Available from: www.fool.com/investing/value/2007/09/rfid-saves-the-dole.aspx

66. D. M. Wasieleski, M. Gal-Or, 2008An enquiry into the ethical efficacy of the use of radio frequency identification technology. Ethics and Information Technology, 1012740

67. A. M. Wicks, J. K. Visich, S. Li, 2006Radio frequency identification applications in healthcare. International Journal of Healthcare Technology and Management, 76522540

68. L. Willey, 2007RFID and consumer privacy: let the buyer beware. Journal of Legal, Ethical and Regulatory Issues, 1022537

69. D. C. Wyld, 2006RFID 101: the next big thing for management. Management Research News, 294154173

70. D. C. Wyld, M. A. Jones, 2007RFID is no fake: the adoption of radio frequency identification technology in the pharmaceutical supply chain. International Journal of Integrated Supply Chain Management, 32156171

71. D. C. Wyld, 2008Radio frequency identification: advanced intelligence for table games in casinos. Cornell Hospitality Quarterly, 492134144

72. F. Zakaria, 2001Time to save "just in time". Newsweek, 1382038

73. Zebra Technologies.2004RFID: the next generation of AIDC. Date of access, September 10, 2006, Available from: www.integratedlabeling.com/rfid/white_papers/Lr2RFIDTechnology.pdf

74. G. A. Zsidisin, 2003Managerial perceptions of supply risk. The Journal of Supply Chain Management, 3911425

75. G. A. Zsidisin, A. Panelli, R. Upton, 2000Purchasing organization involvement in risk assessments, contingency plans, and risk management: an exploratory study. Supply Chain Management: An International Journal, 54187197

76. Y. Zuo, 2010Secure and private search protocols for RFID systems. Information Systems Frontier, 125507519

CHAPTER 3

RFID-Based Vehicle Positioning and Its Applications in Connected Vehicles

Jianqiang Wang [1], Daiheng Ni [2,*] and Keqiang Li [1]

[1] State Key Laboratory of Automotive Safety and Energy, Tsinghua University, Beijing 100084, China

[2] Department of Civil and Environmental Engineering, University of Massachusetts, Amherst, MA 01003, USA

ABSTRACT

This paper proposed an RFID-based vehicle positioning approach to facilitate connected vehicles applications. When a vehicle passes over an RFID tag, the vehicle position is given by the accurate position stored in the tag. At locations without RFID coverage, the vehicle position is estimated from the most recent tag location using a kinematics integration algorithm till updates from the next tag. The accuracy of RFID positioning is verified empirically in two independent ways with one using radar and the other a photoelectric switch. The former is designed to verify whether the dynamic position obtained from RFID tags matches the position measured by radar that is regarded as accurate. The latter aims to verify whether the position estimated from the kinematics integration matches the position obtained from RFID tags. Both means supports the accuracy of RFID-based positioning. As a supplement to GPS which suffers from issues such as inaccuracy and loss of signal, RFID positioning is promising in facilitating connected vehicles applications. Two conceptual applications are provided here with one in vehicle operational control and the other in Level IV intersection control.

KEYWORDS

RFID; vehicle positioning; connected vehicle technology

1. INTRODUCTION

On 3 February 2014, The U.S. Department of Transportation's (DOT) officially announced its decision to move forward with vehicle-to-vehicle communication technology for light vehicles. After decade-long research and experimentation, this decision signifies USDOT's resolution to transform transportation safety and mobility by allowing cars to "talk" with each other. A long list of innovative applications have been tested or under the way including cooperative collision warning [1–3] intersection safety supporting [4], intersection movement assist, *etc*. In these applications, real-time vehicle positioning is assumed in their algorithms and protocols for motion guidance, operational control, and interaction with other vehicles. This is a reasonable assumption since Global Positioning System (GPS) technology has become widely available and affordable. As a matter of fact, many vehicles have already been equipped with GPS for navigation and tracking purposes. However, GPS-based vehicle positioning begins to show its limitations as connected vehicles are advancing toward real-world implementation and especially when the success of these applications depends heavily on the accuracy of vehicle positioning. These limitations include poor or no signals in certain areas especially urban canyon, and positioning accuracy in a dynamic environment.

To address the above limitations, this paper proposes a supplementary yet independent approach, *i.e.*, radio-frequency identification (RFID)-based vehicle positioning, to facilitate connected vehicle applications at critical locations where GPS service is unavailable or unreliable.

This paper is arranged as follows: the next section identifies research gaps based on a survey of literature in terms of vehicle positioning technologies. Following that, Section 3 proposes the RFID-based vehicle positioning approach with design details. Section 4 verifies the accuracy of the RFID-based vehicle positioning in two empirical ways. Section 5 provides two conceptual examples to illustrate the application of RFID-based vehicle positioning in connected vehicles. Lastly, conclusions are drawn in Section 6.

2. RESEARCH GAP IN VEHICLE POSITIONING

Due to its wide coverage and availability, GPS seems to be ideal for connected vehicle applications [5,6]. Stand-alone GPS has the capability of achieving an accuracy of about 20–30 m, which can be narrowed down to 8–12 m after the removal of the selective availability [7]. Differential GPS can even enhance accuracy further to 1–2 m. However, it relies on ground-based reference stations which only covers limited areas and, thus, significantly drives up the cost [8].

In connected vehicles, the requirement on vehicle positioning varies with the nature of application. In general, those involving a large spatial and temporal scope, such as roadway incidence assistance and dynamic routing, do not require accurate positioning updated at a high frequency. For example, an accuracy of 5–10 m is required to warn drivers of hazard at a fixed location (e.g., an accident site), to which general purpose GPS receivers suffice the need. In contrast, applications in small spatial and temporal areas, such as motion control especially crash avoidance, require accurate positioning in real time. For example, most safety applications require one to two meters [9] Shladover *et al.* [10] pointed out that assigning vehicles to the correct lanes would require a standard deviation of about 1 m, but 50 cm accuracy is likely to produce significantly better performance especially for blind spot warning. Combined with adverse locations such as urban canyon, these situations pose a great challenge to GPS-based positioning. As such, further enhancements of GPS or alternative vehicle positioning technologies are called for. Consequently, a number of approaches have been proposed including inertial systems, dead reckoning, information fusion, and map matching.

To improve the positioning performance with GPS/DGPS, a common choice is to integrate it with inertial systems. For example, Farrell *et al.* [11] implemented a real-time carrier phase DGPS aided inertial navigation system which is able to achieve an accuracy at the centimeter level. Huang and Tan [12] used a Kalman filter to incorporate in-vehicle motion sensors in the refinement of vehicle position. However, Jiménez *et al.* [13] pointed out that this approach is valid only if inertial measurements are used before DGPS signals begin to degrade. In addition, the fine accuracy relies on DGPS which is a costly solution only available at limited locations. Dead reckoning [14] advances a vehicle's position from its last known position by integrating its speeds over elapsed time and course. However, this approach is only good for a short period of time and is subject to cumulative errors.

Closely related to the use of inertial systems is the fusion of information [15]. For example, GPS signals can be combined with inertial sensors and digital maps to infer the best estimate of vehicle location [16]; Edelmayer *et al.* [17] used a cooperative federated filtering approach to enhance position estimation based on a variety of position measurements, e.g., from on-board vehicle positioning system, from other cooperating vehicles in the vicinity, as well as from the immediate roadside environment via communication. Bevly [18] attempted to correct inertial sensor errors by using a kinematic Kalman filter estimator to integrate GPS signals, accelerometers, and rate gyroscopes. Islam *et al.* [19] implemented a multi sensor system consisting of a single-axis gyroscope and an odometer integrated with GPS receiver. Though information fusion can achieve high accuracy in some cases, the resultant position is inevitably an estimate that depends on multiple sources of information. Error or missing of a component would degrade estimation quality.

Map matching is to determine the position of a vehicle by constructing a trajectory from a few reliable locations that the vehicle has recently passed and then matching this trajectory to a digital map to find the best fit among multiple likely arcs [20,21]. This approach is best suited for applications that rely on a GPS receiver as the sole means of positioning. However, the uncertainty introduced by inferences in the underlying algorithm limits its use in safety-related applications.

Therefore, in order to obtain high reliability, low cost, and sufficient accuracy under all operational conditions, there exists a great demand for alternative approaches that are readily available, do not rely on GPS, and minimize the need for estimation and fusion. In this context, the approach of using radio sensors such as infrared, microwave, and radio frequency devices has received increasing attention [22,23]. Capable of tracking moving objects [24], these devices can be mounted at roadside to transmit and receive data from vehicles passing in close proximity if they are equipped with transceivers. These systems have been employed in several research projects [25–27] and already been used in transportation such as vehicle speed control [28], real-time bus recognition [29], group location management [30], and electronic toll collection. Due to its low cost and reasonable accuracy, radio-frequency identification (RFID) is promising as a supplement to GPS in connected vehicle applications at critical locations where GPS is unavailable or unreliable but the demand for real-time positioning is high. In the next section, we present an RFID-based vehicle positioning approach and two conceptual applications of the above nature are provided in Section 5.

3. RFID-BASED VEHICLE POSITIONING

RFID tags are a series of passive RFID tags which are fastened on road surface containing position information, e.g., the distance to a reference point, lane number, and direction of travel. When a vehicle passes above an RFID tags, the RFID reader and antenna carried by the vehicle activates the tag and reads in the position information. The layout of the RFID position system is illustrated in Figure 1 and hardware installation pictured in Figure 2. An example design of the format of position information is provided in Table 1.

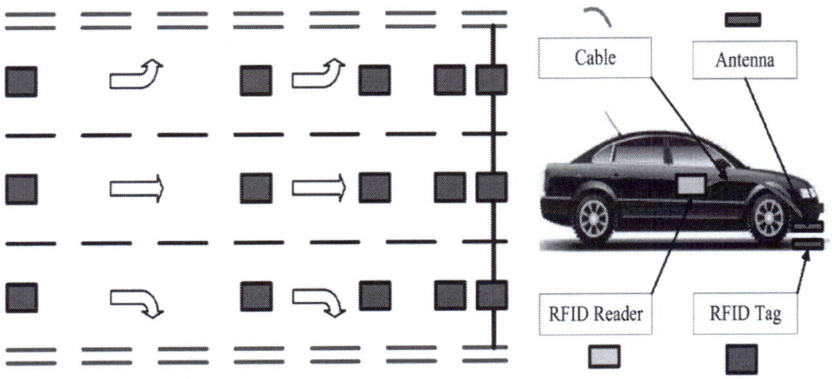

Figure 1. Layout of RFID tags and reader.

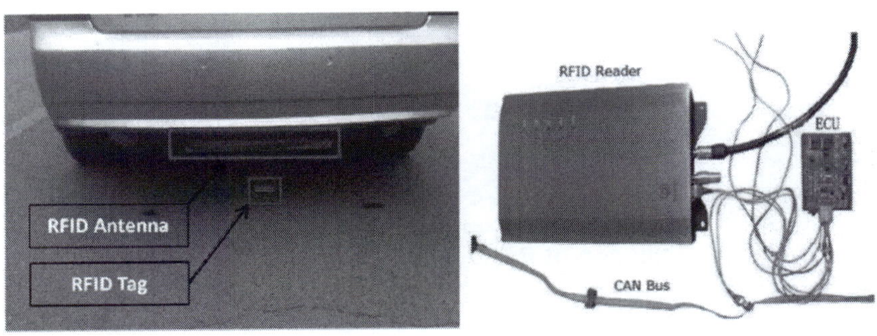

Figure 2. RFID tag, antenna, reader, and controller unit.

Table 1. Definition of the data stored in RFID tags.

Bit	Length	Value Range	Note
0	1	0	Start Bit
1	1	0, 1	Stop Sign: 1: Yes; 0: No
2~9	8	0×00~0×FF	Distance to Stop sign on intersection, Unit: m
10~12	3	0×0~0×7	Lane No.: 0: Go Straight; 1: Turn Left
13~14	2	0~3	2: Turn Right; 3: U-Turn; 4: Go Straight and Turn Left; 5: Go Straight and Turn Right; 6: All Directions
15	1	0, 1	Orientation: 0: East; 1: South;

In this setup, we used an XCAF-12L Panel Antenna (Invengo Information Technology, Co. Ltd., Shenzen, China) which is a rugged UHF directional antenna with a central frequency of 915MHz and circular polarization. The RFID Reader was an Invengo XCRF-502E with a working frequency of 902–928 MHz and working range up to 10 meters. The RFID Tags were ZT-T80s with an effective range of 2–100 m and identification speed up to 200 km/h.

To facilitate the communication between the reader and the tags, an electronic control unit (ECU) was developed. As indicated in Figure 2, the ECU was used to control the reader by RS232 and to transfer data to other modules by CAN Bus. The connection of RFID reader, ECU and CAN Bus is shown in Figure 3. The ECU includes a power module and CPU is based on the Motorola 9s08DZ16 chip. The serial port transceiver is a MAX232 and the CAN transceiver is a TJA1050.

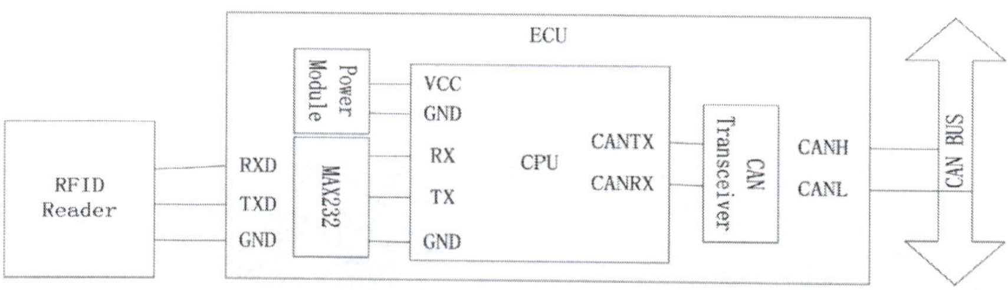

Figure 3. The connection of RFID reader and ECU.

Since each RFID tags contains static position information at a fixed location, a need arises for a vehicle in motion to acquire its accurate position in a continuous fashion in order to support connected vehicle applications. As such, a kinematics integration algorithm has been devised and added to the RFID positioning system, see Figure 4.

The algorithm calculates vehicle position as follows:

$$d_2 = d_{\text{tag}} + d_1$$

$$d_1 = \begin{cases} 0 & F_{\text{tag}} = 1 \\ \sum_k v_k \tau + \frac{1}{2} \sum_k a_k \tau^2 & F_{\text{tag}} = 0 \end{cases}$$

where, d_2 is current position; d_{tag} is the stored position obtained from RFID tag last time; d_1 is estimated driving distance according to speed integral; F_{tag} is a flag whose value is 1 (the system is able to read information from RFID tag) or 0 (otherwise); k is data sequence number, starting to count when the system fails to read the tag and reset to 0 when reading resumes; v and a are vehicle speed and acceleration respectively; τ is time elapsed since last successful reading from RFID tag.

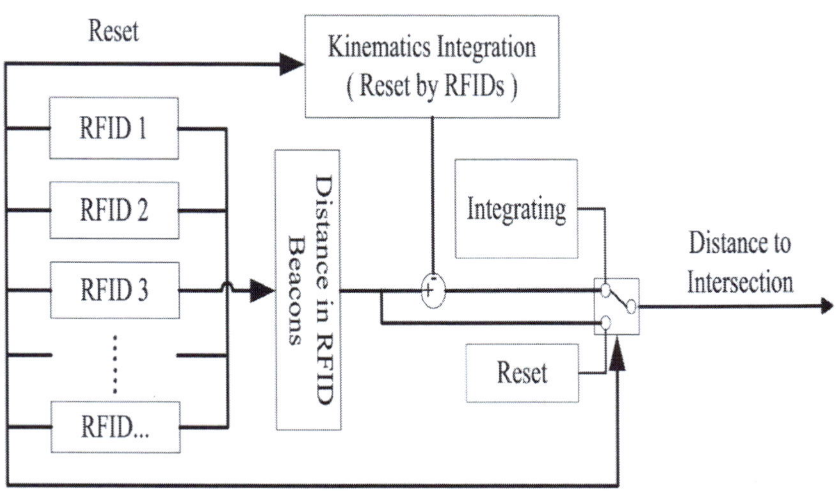

Figure 4. Kinematic integration algorithm.

4. EXPERIMENTAL VERIFICATION

The accuracy of the RFID positioning system can be affected by RFID communication range and distance between tags. Since RFID only communicates within a few meters, reading from a tag only occurs when a vehicle moves over the tag, which ensures accuracy. If the vehicle fails to obtain position updates from tags, its position has to be estimated. The longer the kinematics integration runs, the larger the estimation error. Therefore, it is necessary to avoid long gaps between tags to ensure accuracy. In order to verify the feasibility and the accuracy of the positioning approach, this paper proposes two test methods with one based on radar and the other photoelectric switch.

4.1. Experimental Verification Based on Radar

The objective is to verify whether the dynamic position obtained from the tags matches the "true" position of the vehicle measured by the radar. The experiment is set up as shown in Figure 5. The experiment vehicle is equipped with radar, RFID reader and its antenna. The radar is installed on the vehicle's front fender guard. The radar wave beam is oriented forward in the direction of travel. The antenna is installed below the fender guard, and the surface of the antenna senses the ground. The tags are installed on the test road, at the end of which is a fixed target to help radar measure distance. The radar features a millimeter wave with frequency 76~77 GHz, range up to 180 m, and resolution 0.7 m.

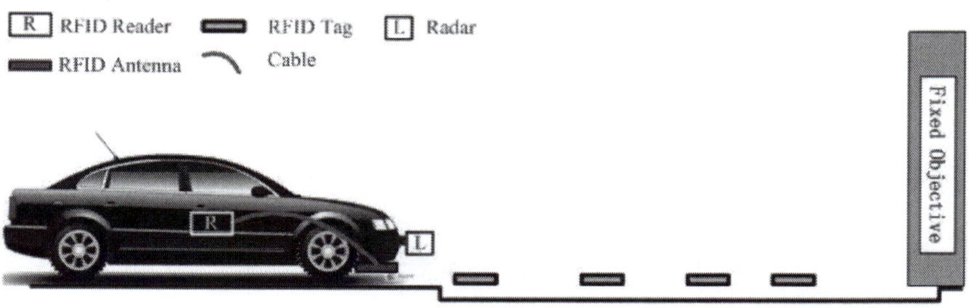

Figure 5. Setup of experimental verification based on radar.

In the experiment, the vehicle passes through each tag consecutively while the vehicle accelerates and decelerates several times. The computer on board calculates the distance between the vehicle and the last tag using the proposed approach. The radar measures the distance between the vehicle and the fixed target independently. The results obtained from these two methods are all transferred to the CAN bus, which can be logged in the computer. The comparison of the test results is shown in Figure 6. Note that the estimated distance is zero at the beginning since there is no tag reading and hence nothing to estimate. Starting from the 5th second, tag readings become available and vehicle position estimation begins. The result shows that positions from radar, tags, and estimation match very well.

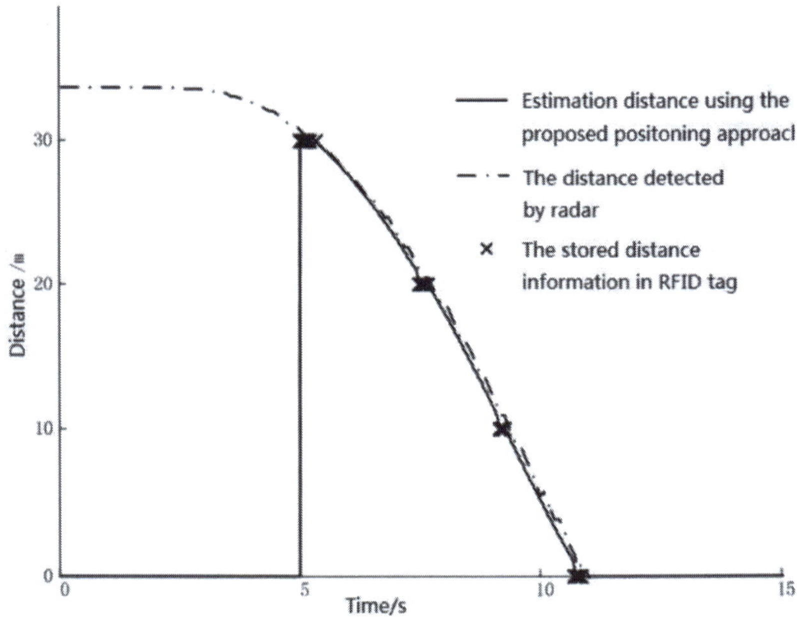

Figure 6. Results of experimental verification based on radar.

4.2. Experimental Verification Based on Photoelectric Switch

The objective focuses on verifying whether the position estimated from kinematics integration matches the position obtained from the tags. The experiment is set up as shown in Figure 7.

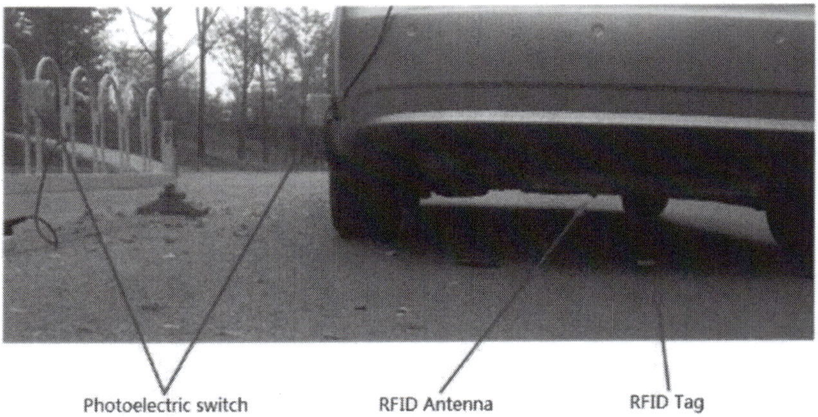

Figure 7. Setup of experimental verification based on photoelectric switch.

The photoelectric switch consists of a transmitter which is fixed at roadside and a receiver which is fixed at the outside of the vehicle. Make sure that the transmitter is in the same cross section as a tag, while the receiver is also in the same cross section as the RFID antenna. As such, when the receiver moves with the vehicle and is aligned with the transmitter, both the RFID and the switch are triggered simultaneously. Starting from this instant, the on-board computer begins to estimate vehicle position using kinematics integration. Meanwhile, another source of position information is obtained from RFID tags.

Figure 8 shows the result of one of the tests. In this test, the error of position is about 5.4% in the first 30 m probably due to accelerating; when speed is relative stable, the errors drops to around 2.5%. It is also noticeable that, as the estimation goes on, the accumulated error increases. Further tests with lower maximum speed (e.g., 36 km/h) reduced the above errors to 3.1% and 1.8% respectively.

The error in position is mainly derived from the accumulation error caused by velocity inaccuracy, especially when the vehicle is accelerating or decelerating. Accordingly, a calibration algorithm is derived using least square method:

$$\Delta d = h_1 + h_2 a$$
$$d_c = \Delta d + d_1$$

where Δd is the position error; d_c is the calibrated position based on integral, a is the vehicle acceleration. The coefficients are estimated as $h_1 = -1.79$ and $h_2 = 0.0613$. After calibration, the errors in the first test drop to 0.07% and 0.66% respectively.

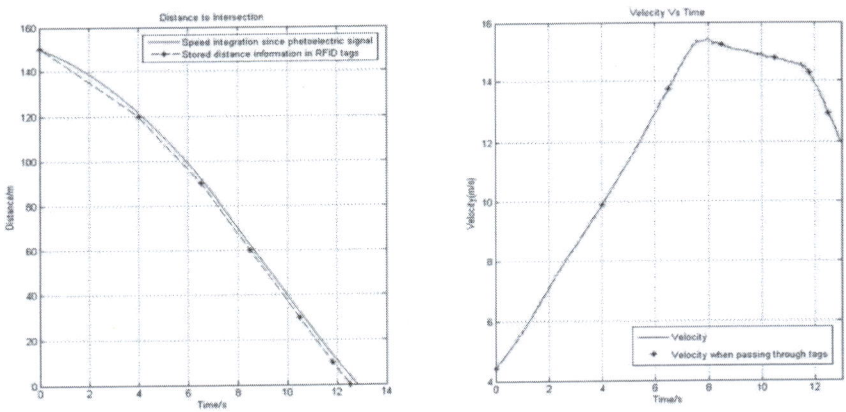

Figure 8. Results of experimental verification based on photoelectric switch.

Limited by time and resources, this research only conducted the above simple, straightforward tests. Nevertheless, the test results revealed that the proposed RFID is promising in providing accurate vehicle positioning in a dynamic process. Before large-scale applications, it is suggested that further tests be performed in more realistic environment (e.g., involving multiple lanes and mixed traffic) with better knowledge of ground truth.

5. EXAMPLE APPLICATIONS IN CONNECTED VEHICLES

Allowing vehicle-to-vehicle and vehicle-to-infrastructure communication, connected vehicle technology opens the door to many innovative applications such as intelligent cruise control [28] that transform safety and throughput. Presented below are two conceptual paradigms in which the above RFID positioning approach helps achieve the goals of connected vehicle technology.

5.1. Vehicle Operational Control

With accurate information about positions and speeds of connected vehicles, it is feasible to synchronize these vehicles on one or more special, managed lanes at high speeds without compromising safety. Such a paradigm is illustrated in Figure 9. RFID tags on the ground pinpoint the location of each vehicle which is equipped with a cooperative driving assistance system.

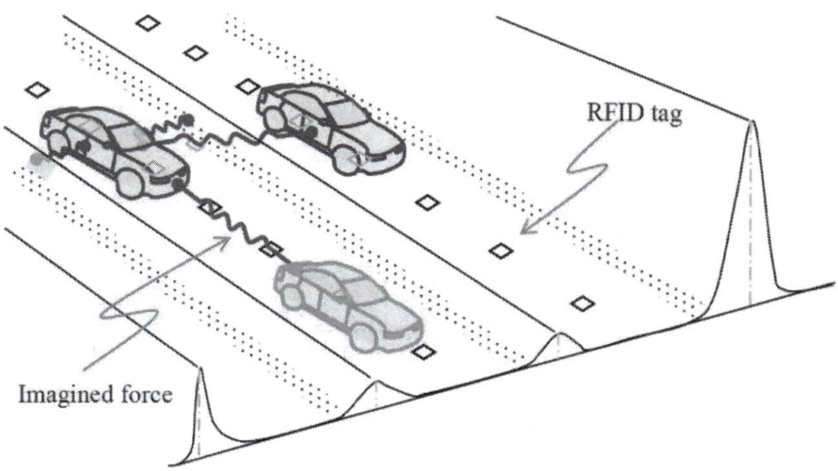

Figure 9. Application of RFID positioning in operational control of connected vehicles.

To ensure safety in the lateral direction, deviation from lane center is translated to a potential field [31,32] that vehicle i needs to overcome. This potential field U i y is imagined in the lateral direction y as bumps along the lane lines, road edge, and center line. By taking the first derivative of U i y with respective to y, one obtains the correction force N_i that is necessary to steer the vehicle back on track:

$$N_i = \partial U_i^y / \partial y$$

where N_i is imagined as the spring between vehicle i and lane line bump, and N_i can be implemented in the actuator that controls vehicle steering. Still in the lateral direction, vehicle k in the vicinity poses a safety hazard. As a result, driver i may choose to "shy" away and this effect becomes more remarkable when k is a heavy truck. Similar to the treatment of lane deviation, the mechanism to avoid parallel running can be created by imagining a repulsive force F i k which is illustrated as the spring between vehicles i and k. Such a force can be derived from the potential field of k perceived by i, U i k :

$$F_i^k = \partial U_i^k / \partial y$$

To ensure safety in the longitudinal direction, a mechanism to maintain safe car following is essential. Hence, the safety hazard in the longitudinal direction x can be represented as a potential field U i j of the leading vehicle j perceived by i. Therefore, the repulsive force F i j that imposes i to keep safe distance can be generically derived from U i j as above, a more concrete form of which can be found in [33]:

$$\ddot{x}_i = A_i \left[1 - \left(\frac{\dot{x}_i}{v_i} \right) - e^{1-\left(\frac{s_{ij}}{s_{ij}^*}\right)} \right]$$

where $\ddot{x}_i$ is the operational control (acceleration or deceleration) of vehicle i, A_i is the maximum acceleration desired by driver i when starting from standing still, $\dot{x}_i$ is the speed of vehicle i, v_i the desired speed of driver i, s_{ij} is the actual spacing between vehicle i and its leading vehicle j, and s i j * is the desired value of s_{ij}.

5.2. Level IV Intersection Control

An intersection is a point in transportation systems where two or more streams of traffic meet and share roadway capacity. To ensure traffic safety, three levels of intersection control are used conventionally. Level I control does not use any physical device to assign priority to traffic, but rather it relies on each driver understanding and observing basic rules specified in Driver's Manuals such as yielding to vehicles on the right and vehicles already in the intersection. If safety hazard poses an issue (typically identified through intersection sight triangle analysis [34,35]), Level II control may be considered which implements YIELD and/or STOP signs to resolve conflict [36]. Currently, the ultimate form of intersection control is Level III, *i.e.*, intersection signalization [36] which alternately assigns right-of-way to specific movements through signal indication such as Green, Yellow, and Red.

Though potentially capable of reducing certain types of crash, Level III control may give rise to other types of collision and negatively impact efficiency. For example, pre-timed signal control ignores the dynamics of approaching traffic, so green time may be wasted on approaches with light or no traffic; Even though actuated control is made traffic-aware, it is not flexible enough to accommodate demands with varying patterns, especially issues caused by unnecessary calls, mandatory minimum green, and arbitrary max out.

Interestingly, the above three levels of control seems to be no match in many aspects to the old-fashioned traffic control by a police officer. For example, the officer is able to watch vehicle clearing an intersection before releasing traffic from a conflicting approach. By clearing before releasing, conflicting vehicles are well protected. In addition, waste of time is minimized since right-of-way is switched right after clearance.

For another example, the officer has full flexibility to assign a relatively long green time to an approach to match its demand or to skip this approach if there is no demand. Moreover, the officer may optimize traffic heuristically on a cycle-by-cycle basis to achieve the overall success of competing objectives such as safety, throughput, and reducing delay.

The only drawback of this officer-directing-traffic paradigm is that it requires the presence of a trained officer around the clock which is impractical. Fortunately, the advent of connected vehicles, combined with sound positioning technology, makes it possible to reproduce this safe yet efficient paradigm electronically which can be called Level IV control. Figure 10 illustrates such a paradigm where each vehicle is able to talk to other vehicles through on-board equipment (OBE) and communicate with the roadside equipment (RSE) at the intersection. The RFID positioning can help by providing real-time, accurate vehicle positions and speeds, with which the RSE can serve as an "electronic police officer" to direct traffic. More specifically, the RSE can send individualized instruction to each driver regarding stop/go and travel speed. Within the RSE, the internal logic dynamically optimizes traffic based on current demands and vehicle positions, resolves conflict, issues customized command to each driver, monitor vehicle status, and update instructions accordingly. Note that the above discussion concerns only about technical feasibility of Level IV control without complicating the problem with legal and moral issues.

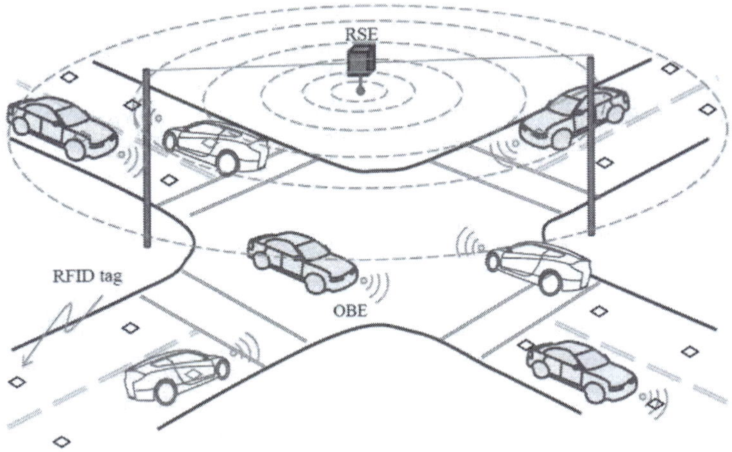

Figure 10. Level IV Intersection control.

6. CONCLUSIONS

This paper proposes an RFID approach as a helpful alternative to positioning in connected vehicle applications where GPS is not available or of poor quality. This approach installs RFID tags on the road surface and on-board tag readers in vehicles. When a reader passes over a tag, the reader can receive the position information stored in the tag. To fill gaps between tags, estimation has to be made based on the latest position update from tags. As such, a kinematics integration method is proposed to serve the purpose. When vehicles accelerate or decelerate, their speeds are changing, which affects the accuracy of the estimation method. Error of this nature can be diminished by applying the proposed calibration algorithm.

Road experiments are carried out to validate the RFID-based positioning approach. One type of experiments involves both radar and RFID reader on board. The radar is used to provide “true” positions of the test vehicle, against which estimates from RFID-based positioning are compared. The result shows good match between the two sources of vehicle positions. The other type of experiments focuses on verifying whether the position estimated from the kinematics integration matches the position obtained from the tags. A photoelectric switch is used to trigger the estimation of vehicle position based on the latest tag update. The results indicate that the error of position is about 5.4% during acceleration or deceleration process and around 2.5% when speed is relative stable. With the help of calibration algorithm, the errors can drop to 0.07% and 0.66% respectively. Before large-scale applications, further tests are recommended in more realistic environment with better knowledge of ground truth.

RFID-based positioning appears promising in connected vehicle applications due to its low cost and reasonable accuracy. Two conceptual applications are conceived in this paper. One application deals with vehicle operation control where RFID position provides accurate vehicle positions to enable the prediction of safety hazard. The other application conceives a Level IV intersection control where RFID position makes it possible to conduct traffic by an “electronic police officer”.

ACKNOWLEDGMENTS

The collaborative research was supported by the National Natural Science Foundation of China, No.: 51175290 and USDOT University Transportation Center (UTC) Program. The authors would like to thank the support of the Chinese National Programs for High Technology Research and Development, No.: 2012AA111901. The authors especially thank Jiaxi Liu and Haipeng Wang (Tsinghua University) for their help with field experiments.

REFERENCES

1. Zhu, H.; Singh, G. A Communication Protocol for a Vehicle Collision Warning System. Proceedings of the 2010 IEEE/ACM International Conference on Green Computing and Communications & 2010 IEEE/ACM International Conference on Cyber, Physical and Social Computing, Hangzhou, China, 18–20 December 2010; pp. 636–644.

2. Tan, H.S.; Huang, J.H. DGPS-Based Vehicle-to-Vehicle Cooperative Collision Warning: Engineering Feasibility Viewpoints. *IEEE Trans. Intell. Transp. Syst.* **2006**, *7*, 415–428.

3. Taleb, T.; Benslimane, A.; Letaief, K.B. Toward an effective risk-conscious and collaborative vehicular collision avoidance system. *IEEE Trans. Veh. Technol.* **2010**, *59*, 1474–1486.

4. Wang, J.; Zhang, D.; Liu, J.; Lu, M.; Li, K. Multi-objective driving assistance system for intersection support. In Proceedings of the 13th International IEEE Conference on Intelligent Transportation Systems (ITSC 2010), Madeira Island, Portugal, 19–22 September 2010; pp. 348–353.

5. Fuersternburg, K.C. A New European Approach for Intersection Safety—The EC-Project INTERSAFE. Proceedings of the 8th International IEEE Conference on Intelligent Transportation Systems, Vienna, Austria, 13–15 Septermber 2005; pp. 343–347.

6. Pierowicz, J.; Jocoy, E.; Lloyd, M.; Bittner, A.; Pirson, B. *Intersection Collision Avoidance Using ITS Countermeasures*; Veridian Engineering Report 8149-12; National Highway Traffic Safety Administration: Washington, DC, USA, 2000.

7. Monteiro, L.; Moore, T.; Hill, C. What is the accuracy of DGPS. *J. Navig.* **2005**, *58*, 207–225.

8. O'Connor, M. Carrier-Phase Differential GPS for Automatic Control of Land Vehicles. Ph.D. Dissertation, Stanford University, Stanford, CA, USA, 1997.

9. Beecroft, M.; Mcdonald, M.; Piao, J.N. Vehicle positioning for improving road safety. *Trans. Rev.* **2010**, *30*, 701–715.

10. Shladover, S.E.; Tan, S.K. Analysis of vehicle positioning accuracy requirements for communication-based cooperative collision warning. *J. Intell. Transp. Sys.* **2006**, *10*, 131–140.

11. Farrell, J.A.; Givargis, T.D.; Barth, M.J. Real-time differential carrier phase GPS-aided INS. *IEEE Trans. Control Sys. Technol.* **2000**, *8*, 709–721.

12. Huang, J.H.; Tan, H.S. A Low-order DGPS-based vehicle positioning system under urban environment. *IEEE/ASME Trans. Mechatron.* **2006**, *11*, 567–575.

13. Jime'nez, F.; Naranjo, J.E.; Garcia, F.; Armingol, J.M. Can low-cost road vehicles positioning systems fulfil accuracy specifications of new ADAS applications? *J. Navig.* **2011**, *64*, 251–264.

14. McLain, P.; Murphy, M.; Vicek, C. GPS/dead reckoning for vehicle tracking in the “urban canyon” environment. Proceedings of the Vehicle Navigation and Information Systems Conference, Ottawa, Canada, 12–15 October 1993; pp. A36–A37.

15. Oliveira, H.A.B.F.; Nakamura, E.F.; Loureiro, A.A.F.; Boukerche, A. Vehicular Ad Hoc Networks: A new challenge for localization-based systems. *Comput. Commun.* **2008**, *31*, 2838–2849.

16. Toledo-Moreo, R.; Zamora-Izquierdo, M.A. Collision avoidance support in roads with lateral and longitudinal maneuver prediction by fusing GPS/IMU and digital maps. *Transp. Res. Part C* **2010**, *18*, 611–625.

17. Edelmayer, A.; Miranda, M.; Nebehaj, V. Cooperative federated filtering approach for enhanced position estimation and sensor fault tolerance in ad-hoc vehicle networks. *IET Intell. Transp. Sys.* **2010**, *4*, 82–92.

18. Bevly, D.M. Global Positioning System (GPS): A Low-Cost Velocity Sensor for Correcting Inertial Sensor Errors on Ground Vehicles. *J. Dyn. Sys. Meas. Control* **2004**, *126*, 255–264.

19. Islam, A.; Iqbal, U.; Langlois, J.M.P.; Noureldin, A. Implementation methodology of embedded land vehicle positioning using an integrated GPS and multi sensor system. *Integr. Comput.-Aided Eng.* **2010**, *17*, 69–83.

20. Srikanthan, T.; Zhang, X.D.; Jagadeesh, G.R. A map matching method for gps based real-time vehicle location. *J. Navig.* **2005**, *57*, 429–440.

21. Bernstein, D.; Kornhauser, A.L.; White, C.E. Some Map Matching Algorithms for Personal Navigation Assistants. *Transp. Res. Part C* **2000**, *8*, 91–108.

22. Kobayashi, M. Vehicle positioning using an infrared beacon and driver behaviour analysis for signal information applied system. Proceedings of the Transportation Research Board Annual Meeting, Washington, DC, USA, 13 January 2008.

23. Mori, S.; Hida, K.; Sawada, K. Advanced car positioning method using infrared beacon. Proceedings of the 8th International Conference on ITS Telecommunications, Hilton Phuket, Thailand, 22–24 October 2008.

24. Han, S.; Kim, J.; Park, C.H.; Yoon, H.C.; Heo, J. Optimal detection range of RFID tag for RFID-based positioning system using the k-NN algorithm. *Sensors* **2009**, *9*, 4543–4558.

25. Farradyne, P.B. *VII Architecture and Functional Requirements (V1.1)*; Report prepared for the ITS Joint Program Office; US Department of Transportation: Washington, DC, USA; 12; April; 2005.

26. Schlingelhof, M.; Betaille, D.; Bonnifait, P.; Demaseure, K. Advanced positioning technologies for co-operative systems. *IET Intell. Transp. Sys.* **2008**, *2*, 81–91.

27. Bohm, M.; Alexander, F.; McDonald, M.; Piao, J. Co-operative infrastructure to vehicle communication to increase road safety along motorways. *J. Eurotransp.* **2007**, *2*, 72–75.

28. Pérez, J.; Seco, F.; Milanés, V.; Jiménez, A.; Díaz, J.C.; Pedro, T.D. An RFID-Based Intelligent Vehicle Speed Controller Using Active Traffic Signals. *Sensors* **2010**, *10*, 5872–5887.

29. Own, C.M.; Lee, D.S.; Wang, T.H.; Wang, D.J.; Ting, Y.L. Performance evaluation of UHF RFID technologies for real-time bus recognition in the Taipei bus station. *Sensors* **2013**, *13*, 7797–7812.

30. Chung, Y.W. Modeling and performance analysis of movement-based group location management using RFID sensing in public transportation systems. *Sensors* **2012**, *12*, 16077–16098.

31. Rossetter, E.J. A Potential Field Framework for Active Vehicle Lankeeping Assistance. PhD Dissertation, Stanford University, Stanford, CA, USA, December 2003.

32. Ni, D. A Unified Perspective on Traffic Flow Theory, Part I: The Field Theory. *Appl. Math. Sci.* **2013**, *7*, 1929–1946.

33. Leonard, J.D.; Jia, C.Q.; Wang, J.Q.; Ni, D.H. Vehicle Longitudinal Control and Traffic Stream Modeling. **2013**. Under review by Transportation Science.

34. AASHTO. *A Policy on Geometric Design of Highways and Streets*, 6th ed.; The American Association of State Highway and Transportation Officials: Washington, DC, USA, 2011.

35. Prassas, E.S.; William, R.; Roger, M.; Roess, P. *Traffic Engineering*, 4th ed.; Prentice Hall: Upper Saddle River, NJ, USA, 2010.

36. FHWA. *Manual on Uniform Traffic Control Devices*; Federal Highway Administration: Washington, DC, USA, 2009.

CHAPTER 4

Investigation of Interference Models for RFID Systems

Linchao Zhang [1], Renato Ferrero [2,*], Filippo Gandino [2] and Maurizio Rebaudengo [2]

[1] China Academy of Electronics and Information Technology, Shuangyuan Road 11, Shijingshan District, Beijing 100041, China

[2] Dipartimento di Automatica e Informatica, Politecnico di Torino, Corso Duca Degli Abruzzi 24, Torino 10129, Italy

ABSTRACT

The reader-to-reader collision in an RFID system is a challenging problem for communications technology. In order to model the interference between RFID readers, different interference models have been proposed, mainly based on two approaches: single and additive interference. The former only considers the interference from one reader within a certain range, whereas the latter takes into account the sum of all of the simultaneous interferences in order to emulate a more realistic behavior. Although the difference between the two approaches has been theoretically analyzed in previous research, their effects on the estimated performance of the reader-to-reader anti-collision protocols have not yet been investigated. In this paper, the influence of the interference model on the anti-collision protocols is studied by simulating a representative state-of-the-art protocol. The results presented in this paper highlight that the use of additive models, although more computationally intensive, is mandatory to improve the performance of anti-collision protocols.

KEYWORDS

RFID; interference model; reader-to-reader collision

1. INTRODUCTION

Radio frequency identification (RFID) is currently a leading technology in the automatic identification and data capture area. With the growing interests of end users in different applications, RFID systems have been exploited in various areas, including smart shopping [1,2], agri-food traceability [3,4], supply chain management [5], construction industry [6], healthcare [7] and temperature tracking for food [8]. Generally, an RFID system is able to increase the process efficiency, by achieving complete traceability and better quality control at a low cost.

A basic RFID system is composed of several tags, one or more readers and a central server. The tag contains data that can be retrieved by a reader located nearby. The server receives, processes and stores the data sent by the readers. Since wireless communication is adopted in an RFID system and the majority of RFID-based applications require a dense reader deployment, data transmissions are susceptible to suffer from interference due to the signals simultaneously broadcast by other readers and tags. The interference in RFID systems can be classified into three types: tag-to-tag, reader-to-tag and reader-to-reader interference [9,10]. Tag-to-tag interference occurs when a reader interrogates multiple tags at the same time. Since this issue affects the performance of all of the RFID systems that involve more than one tag, many approaches have been proposed to address the tag-to-tag collision [11]. Reader-to-tag interference occurs when two or more readers interrogate the same tag simultaneously; this can be solved by separating the interrogation range of the readers. Reader-to-reader interference happens when a reader that is interrogating a tag receives strong signals from one or more readers operating at the same radio frequency, disturbing the weak signal backscattered from the target tag.

So far, although a number of research works related to analyzing reader-to-reader interference have been conducted, there is not an agreed interference model that can address the reader-to-reader collision problem efficiently and accurately. The existing interference models can be roughly categorized into two groups: single interference and additive interference models. In a single interference model, only the collisions between two readers are evaluated. On the contrary, also the interferences caused by a group of readers are considered in an additive interference model: although each reader in the group cannot disturb a certain reader r if considered alone, the interference generated by the whole group would prevent r from identifying tags.

In this paper, the single and additive interference models are evaluated and compared by simulating an RFID system. On the one hand, the single interference models can be simulated more easily, so they are preferred for a faster evaluation. On the other hand, the additive interference models consider a wider set of possible collisions, in a more coherent way with the real behavior of an RFID network. The goal of the paper is to evaluate if the difference in accuracy between the models are so relevant to recommend the use of an additive interference model for a realistic simulation, despite its computational overhead. In order to emulate the behavior of different RFID applications, three scenarios have been analyzed, and a wide range of system configurations has been considered for each of them. The first scenario consists of a set of independent readers: each reader periodically interrogates the tags with a given probability, and its activity is not affected by the other readers. Without any regulation among the readers, the reader-to-reader interference is more likely to occur, thus preventing the tag identification. In order to increase the performance of the RFID network, the adoption of a channel access method is suggested. Time division multiple access (TDMA) [12] is a well-known channel access scheme for shared medium networks. In a TDMA anti-collision protocol, the readers are scheduled to operate at different time slots. The second scenario considers a basic implementation of the TDMA scheme, with a random choice of the time slot for every reader. Many TDMA protocols have been specifically designed for RFID systems. In particular, the third scenario considers that the readers are scheduled according to the distributed color selection protocol (DCS) [13], which is an ad hoc TDMA protocol for RFID networks. The study of the behavior of the RFID system with the three proposed interrogation scenarios allows one to identify two main issues. Firstly, without applying any anti-collision protocol, the reader-to-reader interference dramatically reduces the system performance, i.e., the throughput and the interrogation efficiency. Secondly, the interference model used to simulate a protocol notably affects the estimated performance. Therefore, a precise evaluation of the reader-to-reader anti-collision protocols requires a proper selection of the interference model: in this way, the configuration parameters can be set more suitably in order to obtain a better performance.

The remainder of the paper proceeds as follows. The most common interference models are reviewed in Section 2. Section 3 describes the evaluation parameters. The simulation results under the three scenarios are presented in **Section 4, Section 5 and Section 6. Finally, Section 7 is devoted to the conclusions.**

2. RELATED WORK

This section reviews the most common models that are applied to describe the interference in an RFID network. The models are categorized according to the type of reader-to-reader collision. Collisions that involve only two readers are called direct collisions: they represent the only source of interference assumed by single interference models. Instead, collisions due to multiple readers are called additive collisions and are taken into account by additive interference models. A reader r experiences an additive collision if there are no other readers causing a direct collision; however, r still cannot identify the tags because other farther readers generate a combined interference.

2.1. Single Interference Models

The single interference model represents the direct collisions that occur in the RFID network by means of an interference graph. Each reader is assigned to a node, and two nodes are linked by an edge if the corresponding readers, referred to as neighbors, may collide when they query tags at the same time. The main interference graphs are described in the following.

2.1.1. Bernoulli Random Graph

The Bernoulli random graph is the simplest interference model. Given a set of nodes, the edges are randomly added. The kind of stochastic process applied to build the graph leads to different variants of the Bernoulli random graph. For example, for every pair of nodes, an edge may be added with a probability p_{Brg} independent from every other edge [14]. Another common approach for building a Bernoulli random graph is defined in [15]. The first step is to fix the total number of edges N_{Brg} in the graph. Then, given a set of vertices, all of the possible graphs with N_{Brg} edges are generated. Finally, one member of that set is randomly and uniformly chosen as the Bernoulli random graph.

The Bernoulli random graph models the randomness of the signal propagation in RFID networks due to external causes. Its main advantage is the simplicity of the model, which enables one to exactly express many average properties of the network. However, it does not properly capture the behavior of real-world networks, because the probability of a link between two nodes is independent of their real distance.

2.1.2. Unit Disk Graph

A collision between two RFID readers can be simply predicted by evaluating their distance: the readers may collide if their distance is lower than a threshold D. Since the only condition for generating a collision is the relative distance between two readers, two nodes r_1 and

r_2 are linked by an edge if r_1 is located within a circle centered in r_2 and with radius D. This interference graph is called the unit disk graph [16].

Some variants of the basic model exist. In a random geometric graph, the position of the readers is randomly chosen inside the unit square $[0,1]^2$ [17]. This interference graph becomes a random sector graphs [18] if the antenna of the readers is directional.

Threshold D can be evaluated in different ways. An explicit formula is obtained in [19] considering a homogeneous RFID network. In that case, the interrogation range d, the transmit power P_r and the antenna gain G_r are the same for all of the readers. Similarly, the antenna gain G_t and the power reflection coefficient R_t are the same for all of the tags. Under the condition of a homogeneous network, the value of the threshold is:

$$D = \sqrt[\alpha]{\frac{K_0 \Gamma P_r d^{2\alpha} G_r^2}{R_t P_r G_t^2 G_r^2 - N_0 K_0^2 \Gamma d^{2\alpha}}} \quad (1)$$

where α is the path loss exponent, Γ is the SINR (signal to interference plus noise ratio) needed to identify the tag, N_0 is the background noise power and K_0 is a coefficient that depends on the adopted propagation model (for example, $K_0 = (4\pi\lambda)^2$ in the free space model).

The analysis in [20] determines the presence of a reader-to-reader collision by correlating the distances between the two readers and the group of tags to query. Reader r_1 can identify tags near it without any interference from reader r_2 if the following condition is satisfied:

$$|r_1 - r_2| \geq (1 + \Delta)|r_1 - t| \quad (2)$$

where Δ is a positive constant, $|r_1 - r_2|$ is the distance between the two readers and $|r_1 - t|$ is the distance between r_1 and the farthest tag t that is queried. Since the longest distance at which a tag can be identified corresponds to the interrogation range d, the longest distance expected in Equation (2) such that two readers experience a reader-to-reader collision is:

$$D = (1 + \Delta)d \tag{3}$$

Equation (3) can be simplified by assigning a constant value to D [21], without any explicit correlation to the interrogation range.

2.1.3. Quasi Unit Disk Graph

Ideally, with an omnidirectional antenna, the interference area generated by a reader is circular, because its emitted signals propagate in all directions in the same way. However, this does not hold up in a real scenario, for example due to the irregular shape of the field, some obstacles that obstruct the signal propagation, the atmospheric conditions, etc. The quasi unit disk graphs models this situation by introducing a parameter ρ, with $0<\rho<1$ [22]. As in the unit disk graph, two nodes farther than D do not interfere. However, a distance lower than D does not entail an interference between the nodes. In detail, in the quasi unit disk graph, two nodes are surely linked only if their distance is lower than ρD . If the distance ranges between ρD and D, the existence of the edge is unspecified.

A natural extension of the quasi unit disk graph is the quasi random geometric graph [23], where the nodes are independent and uniformly distributed on the unit square [0,1] 2 . The presence of obstacles that reduce the interference between a pair of readers can be captured by other graphs, such as the faulty random geometric graph [24], which is obtained from a random geometric graph through random removal of edges and vertices.

2.2. Additive Interference Models

The criterion followed by the additive interference models for determining the occurrence of a reader-to-reader collision is the evaluation of the SINR at the reader's antenna. Let be $P_{t,r}$ the power of the reply that reader r receives from tag t and I the interference generated by the other readers. r can identify t only if the measured SINR is higher than a threshold Γ:

$$\frac{P_{t,r}}{I + N_0} \geq \Gamma \tag{4}$$

The contributions to I due to each reader can be represented with a weighted undirected graph: the weight $w_{i,0}$ of an edge between two readers r_i and r_0 expresses the interference that the signal emitted by r_i provokes on r_0 . Reader r_0 experiences a reader-to-reader collision if the sum of the weights of all its incoming edges in the graph is so high that Equation (4) is not satisfied:

$$\frac{P_{t,r}}{\sum_{i=1}^{N-1} w_{i,0} + N_0} < \Gamma \tag{5}$$

The following subsections present two categories of additive interference models, which differ in the quantity of readers that are considered to compute the interference perceived by the target reader.

2.2.1. Complete Weighted Graph

The most general additive interference models compute the interference received by reader r_0 by summing the contributions of all of the other readers:

$$I_0 = \sum_{i=1}^{N-1} P_{r_i,r_0} \tag{6}$$

where N is the number of the readers in the network and P_{r_i,r_0} is the power of the signal transmitted by r_i, measured at r_0's antenna. The resulting interference graph corresponds to a complete weighted graph: each node is linked to all of the others, and an edge between two nodes r_i and r_j has weight P_{r_i,r_j}.

Different approaches for estimating the power of the signals $P_{t,r}$ in Equation (4) and P_{r_i,r_j} in Equation (6) exist [19]. A reader can only query tags closer than d, so the lowest power of the signal received by a tag is:

$$P_{r,t} = P_r \frac{G_r G_t}{K_0 d^{\alpha}} \tag{7}$$

Since only a fraction of the signal is backscattered by the tag, the lowest power of the tag's reply received by the reader is:

$$P_{t,r} = R_t P_{r,t} \frac{G_t G_r}{K_0 d^{\alpha}} \tag{8}$$

In a homogeneous network, r_1 generates an interference on the transmission of r_0 equal to:

$$P_{r_1,r_0} = P_r \frac{G_r G_r}{K_0 |r_1 - r_0|^{\alpha}} \tag{9}$$

thus, the amount of interference perceived by r_0 is:

$$I_0 = \sum_{i=1}^{N-1} P_r \frac{G_r G_r}{K_0 |r_i - r_0|^\alpha} \tag{10}$$

2.2.2. Random Proximity Graph

The interference caused by a reader on another one is inversely proportional to their distance, as shown in Equation (9), so the greatest amount of interference is due to close readers. Therefore, the total interference can be satisfactorily approximated by summing the contributions of the k closest readers. The obtained interference graph is called the random proximity graph [25]. In this graph, the deployment area is the unit square [0,1] 2 , and the nodes are randomly located with a uniform distribution. Each node is linked with its k closest nodes, which are identified according to the Euclidean distance.

The random proximity graph can be generalized to a K-nearest neighbor graph [26] if a metric space other than the Euclidean space is considered. In that case, the number of dimensions is higher than two, and a measure of distance different from the Euclidean distance may be applied.

3. SIMULATION SETUP

An ad hoc simulator for RFID networks [27] has been adopted to test and compare the single and additive interference models. In particular, the unit disk graph and the complete weighted graph models are chosen as representative of the two classes of models.

In the experiments, the deployment area is a 1000 m × 1000 m field. In order to evaluate the influence of the reader density, the number of readers varies from 10 to 50. The RFID network is considered homogeneous. Table 1 shows the values of the parameters adopted in the simulation, in accordance with [28]. In particular, D is computed by substituting the adopted values into Equation (1). D is used to determine whether two readers are neighbors in the unit disk graph model.

The following metrics have been adopted to compare the interference models:

success rate: the ratio of the number of successful interrogations to the amount of attempts;

throughput: the average number of successful interrogations performed per reader during a specific amount of time;

percentage of additive collisions (for the additive interference model only): the ratio of the number of collisions caused by more than one reader to the amount of reader-to-reader collisions.

Table 1. Parameters of the interference models.

Parameter	Symbol	Value
Interrogation range	d	5 m
Reader antenna gain	G_r	6 dBi
Tag antenna gain	G_t	1 dBi
Tag's power reflection coefficient	R_t	3/4
Reader's transmit power	P_r	30 dBm
Model coefficient	K_0	$1/G_r^2$
Path loss exponent	α	2
SINR Threshold	Γ	10
Noise power	N_0	0
Threshold distance	D	288.68 m

As the activity of the readers depends on the application requirements, different behaviors are tested by considering three scenarios in the simulations:

probabilistic interrogation: at every time slot, each reader has a probability p of querying tags;

slotted interrogation: the communication is organized in rounds composed by a fixed number of time slots, and each reader randomly chooses a time slot to interrogate the tags;

DCS protocol: the readers are scheduled according to a state-of-the-art TDMA protocol [13].

Different configuration parameters have been considered for every scenario: each case is simulated 100 times in order to reduce the effect of randomness.

4. PROBABILISTIC INTERROGATION SCENARIO

In the current scenario, the activity of the readers is modeled as a Bernoulli process, i.e., a discrete-time stochastic process. The simulation time is divided into iterations of the same length: in the experiments, each simulation consists of 2000 iterations. The last instant of an iteration is called the interrogation point. The readers are allowed to interrogate the tags only at the interrogation points, with a probability p. The probability p, which represents the interrogation frequency of the system, is selected according to the goal of the system: the higher p, the more often the tags are queried. The activity of a reader at the interrogation point is independent of the other readers. Furthermore, the operation is memoryless, i.e., the behavior of a reader is independent of the previous iterations.

The throughput of the readers depends on p and on the density of the nodes: for example, few tag identifications are expected to be completed with a low density and a low value of p (because few interrogations are attempted) or with a high density and a high value of p (because there is too much interference generated by the simultaneous interrogations). Different simulations of the probabilistic interrogation scenario are performed, with p ranging from 0.1 to 1.

4.1. Success Rate in the Probabilistic Interrogation Scenario

Figure 1 compares the success rate in the two models, with 30 or 50 readers in the network. It can be observed that the success rate is always lower than 0.6 for 30 readers and 0.4 for 50 readers: this indicates that an anti-collision algorithm is necessary to reach a good probability of tag identification in an RFID system. Beyond a certain threshold of p, called bottleneck probability, it is impossible to perform any tag identification because the quantity of readers simultaneously querying is too high. As shown in Figure 1a, with 30 readers, the bottleneck probability is 0.8 according to the single interference model, but it falls to 0.4 according to the additive interference model. Similarly, as shown in Figure 1b, with 50 readers, the bottleneck probability reduces from 0.6 to 0.2, by passing from the single to the additive interference model. This remarkable difference in the bottleneck probability reveals to what extent the behavior of the readers differs according to the single and additive interference models.

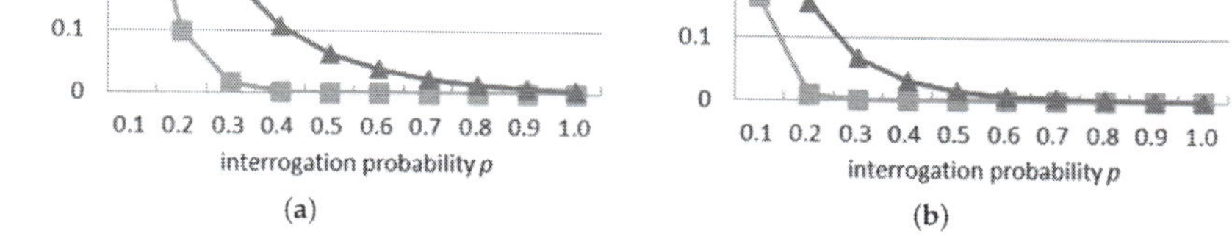

Figure 1. Success rate in the probabilistic interrogation scenario: (a) 30 readers; (b) 50 readers.

In both the networks, the successful ratio results always in being higher by using the additive interference model than by using the single interference one, since the additive interference model is able to consider a larger number of reader-to-reader collisions with respect to the single interference one.

The success rate in Figure 1b is always lower than the corresponding rate in Figure 1a, because, as the network density grows, the interrogations of a reader influence more neighbors.

4.2. Throughput in the Probabilistic Interrogation Scenario

Figure 2a considers the probabilistic interrogation scenario and shows the throughput computed by using the single interference model. Considering a quantity of readers between 10 and 30, initially, the throughput increases, but after a peak, it decreases. The position of the peak is due to p: as the quantity of readers increases, the peak is reached with a lower probability. When there are more than 30 readers, the throughput constantly decreases as p grows. For a fixed interrogation probability, the throughput decreases if the quantity of readers grows, because a larger number of readers involves a greater number of simultaneous interferences.

Figure 2b presents the throughput expected by the additive interference model. The peak point is reached with a value of p higher than 0.1 only if there are 10 readers or less in the network. In accordance with the single interference model, the throughput decreases if the quantity of readers grows due to the bigger amount of collisions among them. It can be also observed that the throughput quickly becomes zero in some system configurations (e.g., 30 readers with $p > 0.4$), in accordance with the success rate shown in Figure 1.

By comparing Figure 2a,b, it can be noted that the value of p associated with the peak point, which represents the best configuration, notably differs according to the two interference models when there are 30 readers or less. Instead, the two models agree with more than 40 readers: in this case, the peak point is always reached with $p=0.1$.

Independently of the number of readers, the highest throughput achievable according to the single interference model is higher than the throughput estimated by using the additive interference one. Figure 3 compares the best throughput according to the two interference models considering a quantity of readers between 10 and 50. It can be observed that the absolute difference in throughput decreases if the quantity of readers grows. However the relative difference increases. In fact, the percentage difference, i.e., the difference between the two values divided by their average, rises from 39% with 10 readers to 45% with 50 readers.

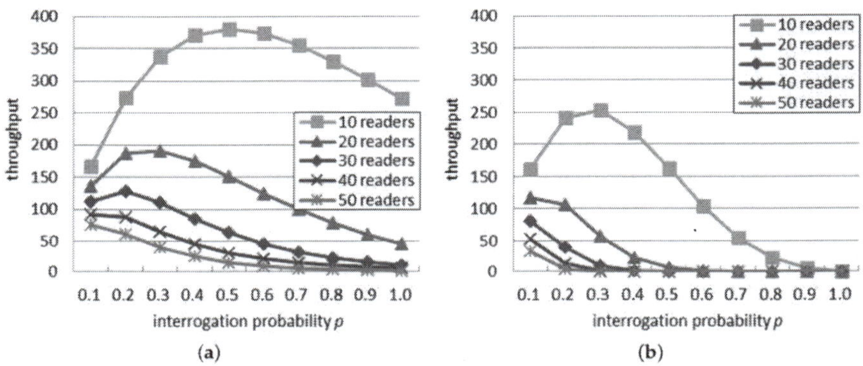

Figure 2. Throughput in the probabilistic interrogation scenario. (a) Single interference model; (b) Additive interference model.

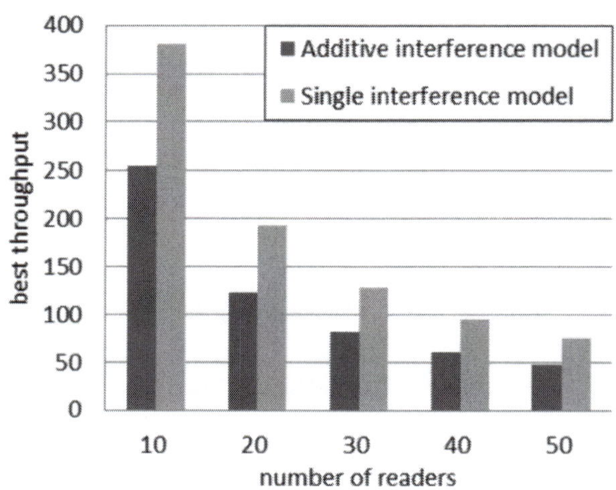

Figure 3. Best throughput in the probabilistic interrogation scenario.

4.3. Percentage of Additive Collisions in the Probabilistic Interrogation Scenario

Figure 4 shows the ratio of collisions due to an additive interference. When the number of readers is between 10 and 30, the ratio initially grows as the interrogation probability grows, because the number of collisions is limited and the relative quota of the additive ones is significant. However, after reaching a peak point, the percentage of additive collisions starts to decrease as p increases, because the direct collisions become predominant: in this case, it is more likely for a reader to have a neighbor that tries to query the tags nearby, causing a collision. With 40 or more readers, the ratio of additive collisions over the total quantity of collisions decreases with a progressively lower impact on the total interference. Independently of the number of readers, the greatest ratio of additive collisions over the total quantity of collisions is about 22%.

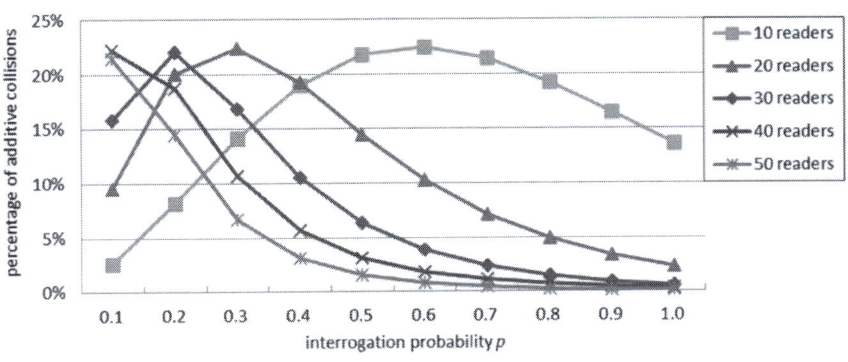

Figure 4. Percentage of additive collisions for the additive interference model in the probabilistic interrogation scenario.

5. SLOTTED INTERROGATION SCENARIO

In this scenario, the communications are organized in rounds with the same duration. The rounds are composed by time slots. Each reader randomly selects a time slot to query tags. If there is a collision, the reader randomly picks a new slot for the further rounds. If the quantity of time slots is high, after a self adjustment phase, the system becomes steady, and the readers are able to query tags without reader-to-reader collisions.

This slotted mechanism is at the base of any TDMA protocol, because it reduces the reader-to-reader collisions. Nevertheless, the quantity of slots per round may affect the performance. In fact, a large quantity of time slots does not correspond directly to a high throughput, because if there are too many time slots, some of them could remain unused. In this section, the trade-off between the throughput and the number of collisions is evaluated trough simulations. All of the simulations last 1000 s, and the length of the time slot is set to 0.5 s.

5.1. Success Rate in the Slotted Interrogation Scenario

Figure 5 illustrates the success rate according to the two interference models. The rate monotonically climbs up as the number of time slots increases, because the probability that more than one reader transmits at the same time slot reduces. The highest number of time slots adopted in the simulations is 30. In this case, with 50 readers, the success rate is equal to 0.63 according to the additive interference model, and it is even higher, equal to 0.71, according to the single interference model.

With respect to the results obtained in the probabilistic interrogation, the success rate is remarkable higher, since the interrogations are scheduled in different time slots, and this reduces the reader-to-reader collisions. However, Figure 5a shows that if the quantity of time slots and of readers are the same, the success rate is lower than one. This result is due to the starting unsteady phase, during which the readers look for a free time slot.

Instead, the difference between the two models is primarily influenced by the quantity of slots: it reduces as the quantity of slots increases. With a larger quantity of time slots, the readers have a greater probability to be assigned to a different time slot, with negligible interference among them.

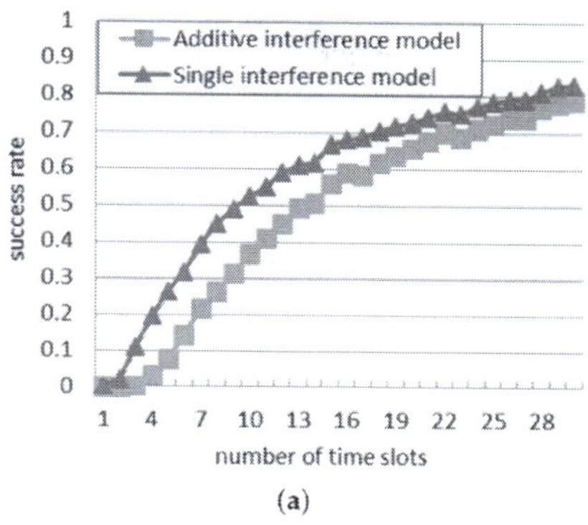

Figure 5. Success rate in the slotted interrogation scenario: (a) 30 readers; (b) 50 readers.

5.2. Throughput in the Slotted Interrogation Scenario

Figure 6 shows the throughput estimated by the two interference models. The general behavior is similar in both the interference models and with any network density: the throughput raises to a peak and then slowly decreases. The position of the peak changes according to the number of readers. In both models, a large quantity of readers corresponds to many time slots needed to obtain the best throughput.

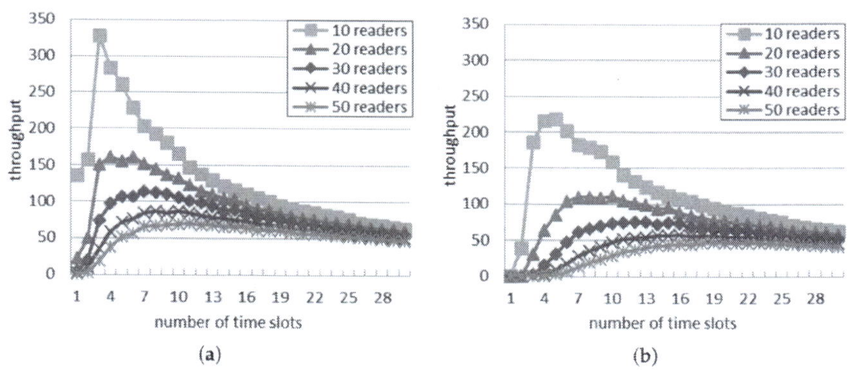

Figure 6. Throughput in the slotted interrogation scenario. (a) Single interference model; (b) Additive interference model.

The throughput predicted by the single interference model is always larger than the throughput computed with the additive model, while the quantity of slots corresponding to the best throughput is lower for the single interference model (e.g., with 10 readers, five time slots per round correspond to the best throughput in Figure 6b; however, in Figure 6a, the supposed best number of time slots is three). The difference is greater for a large quantity of readers: with 50 readers, the best quantity of slots per round is 11 and 19 for the single and additive interference models, respectively.

Figure 7 shows the comparison of the best throughput with the considered models. The percentage difference between the best throughput supposed by the considered models is almost constant and independent of the number of readers. It oscillates in the interval between 37% and 40%. For a larger quantity of readers, the best throughput is achieved by using a higher quantity of slots, as shown in Figure 6. As a consequence, the percentage difference between the best throughput in the two models is not affected by the number of readers, because it is compensated by using a different quantity of slots.

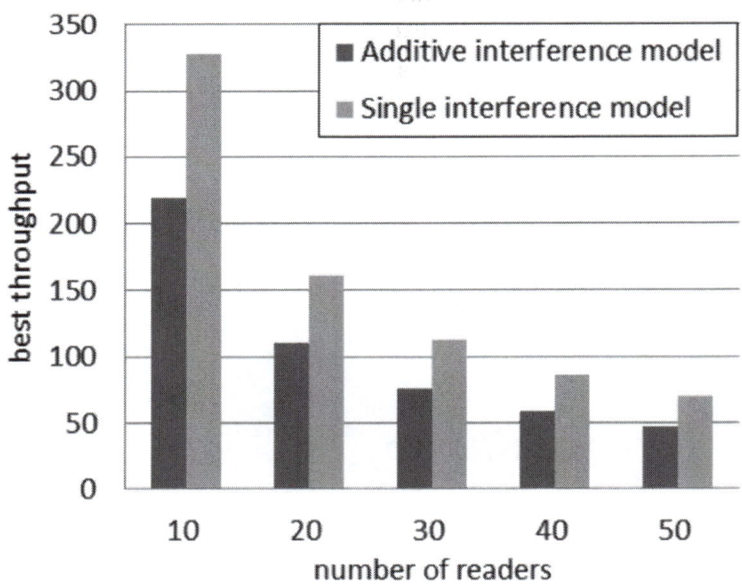

Figure 7. Best throughput in the slotted interrogation scenario.

5.3. Percentage of Additive Collisions in the Slotted Interrogation Scenario

Figure 8 considers the scenario of slotted interrogation and shows the ratio of additive collisions over the total quantity of collisions. Only with 10 readers, the ratio of additive collisions is affected by the number of slots per round. This confirms the trend of Figure 7: by varying the number of readers, the best throughput is achieved by using a different quantity of slots. Since the percentage of additive collisions is poorly affected by the quantity of slots, the percentage difference is similar with both models.

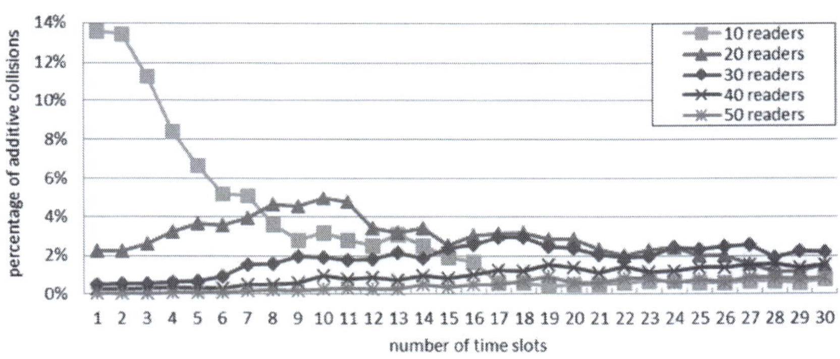

Figure 8. Ratio of additive collisions with the slotted interrogation scenario.

6. DCS SCENARIO

The distributed color selection (DCS) protocol [13] is one of the most popular TDMA anti-collision protocols. The main idea is to use a certain number of colors (maxColors) to distinguish all of the readers in the system, so that each reader has the smallest possible number of neighbors with the same color. Initially, each reader selects a color and queries the tags only during the corresponding time slot. Once a reader detects an interrogation failure, it randomly changes the color and broadcasts the new color to its neighbors: this message is referred to as a kick.

Whenever a reader receives a kick with the same color as its own, it selects another color.

With respect to the previous scenario, in DCS, the choice of a new time slot for a reader may be due not only to its own activity (i.e., after an interrogation failure), but also to the information received from its neighbors (i.e., when they send a kick). However, no countermeasures are planned in DCS if a reader experiences an interference generated by a group of other readers. This event is considered only by the additive interference model, so the performance of DCS is expected to change depending on the considered interference model. The goal of this section is to investigate the different performance of the DCS protocol according to the single and additive interference models. The length of each slot is 0.5 s, and the total simulation time is 1000 s.

6.1. Success Rate in the DCS Scenario

Figure 9 presents the success rate achieved with DCS according to the two interference models. The values reported in both models decrease as more readers are deployed, since more collisions appear when the density grows. When there are 30 readers, DCS shows a good performance (i.e., the success rate is larger than 99%) with maxColors≥ 10 in the single interference model and maxColors≥ 15 in the additive interference model. However, when the number of readers grows to 50, a big raise of maxColors is required in order to get the same performance: the values of maxColors required to reach a success rate higher than 99% become 18 and 25 for the single and additive interference models, respectively.

It can be observed that the difference between the two models is much larger than in the case of probabilistic and slotted interrogations. For example, in Figure 9a, if the value of maxColors is lower than or equal to three, the success rate is close to zero according to the additive interference model: this means that three colors are not enough to avoid collisions in such a dense deployment. On the contrary, with three colors, the single interference model supposes that 18% of the interrogations are successful. According to the additive interference model, a similar success rate is achieved only if maxColors is at least equal to eight. Furthermore, the difference between the considered interference models is larger for a greater quantity of readers. In Figure 9b, the additive interference model estimates a success rate close to zero with a value of maxColors lower than or equal to six: this configuration leads to a success rate of 24% in the single interference model. maxColors should be at least equal to 15 to achieve such a success rate in the additive interference model. This analysis reveals that the choice of the interference model plays an important role in the design of an anti-collision protocol, in order to set the proper configuration parameters (like the value of maxColors in DCS depending on the network density).

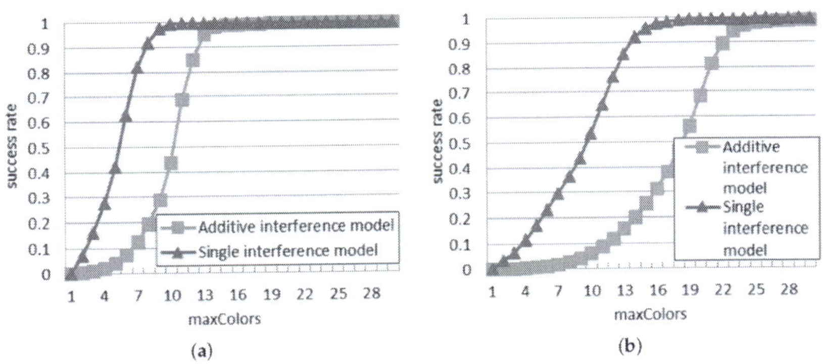

Figure 9. Success rate in the distributed color selection (DCS) scenario: (a) 30 readers; (b) 50 readers.

6.2. Throughput in the DCS Scenario

Figure 10 presents the throughput of the DCS protocol in the single and additive interference models. On the one hand, a large value of maxColors results in a good ability to avoid collisions, but on the other hand, it also means fewer interrogation attempts at the same time. This explains the trend of the throughput in both the inference models: initially, the throughput grows as maxColors increases, because the number of reader-to-reader collisions reduces. However, after reaching a peak, the throughput decreases if maxColors further increases, because the number of reader-to-reader collisions keeps stable, but the readers query tags less frequently.

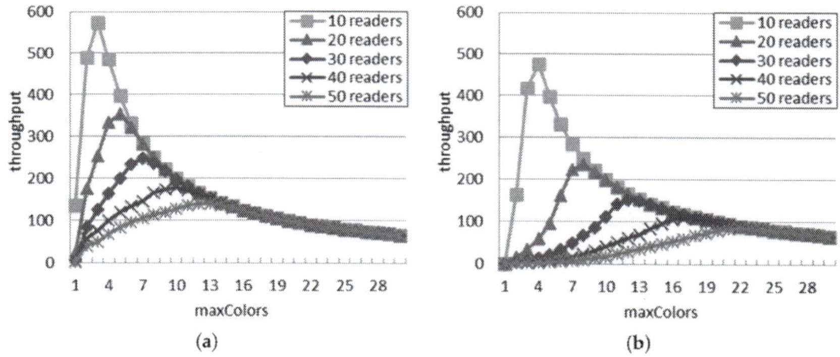

Figure 10. Throughput in the DCS scenario. (a) Single interference model; (b) Additive interference model.

The single interference model tends to overestimate the performance of the DCS protocol, since it ignores the additive collisions. Besides the throughput loss, the additive collisions may affect the behavior of the network in other ways: first, more kick packets are exchanged in the additive interference model because a reader can collide also with a group of readers and not only with a single one, so it needs to change color more often; secondly, it is more likely that its new color will cause another collision, because, besides the colors of its neighbors, the reader should avoid also the colors of readers that can generate an additive collision.

6.3. Best Configuration in the DCS Scenario

As shown in Table 2, the optimal value of maxColors in order to achieve the highest throughput depends on the interference model. For example, in a network with 40 readers, maxColors should be 10 according to the single interference model, but that value turns to 17 according to the additive interference model. The difference between that optimal value grows as the quantity of readers rises: for example, the difference raises from one, with 10 readers, to 10, with 50 readers. Considering that the additive model is more accurate, it can be concluded that the single interference model is not able to compute the optimal parameter setting for the DCS protocol, especially if there are many readers. The fourth and the fifth columns in Table 2 report the throughput according to the additive interference model for two distinct values of maxColors : the value suggested by the single interference model (as shown in the second column) and the value suggested by the additive interference model (as shown in the third column). If there are more than 10 readers, by selecting the value of maxColors according to the single interference model, the throughput loss, with respect to the configuration selected according to the additive interference model, ranges from 59.4% to 67.4%. The difference is less significant with 10 readers because the optimal value of maxColors suggested by the two interference models does not differ significantly.

Table 2. Analysis of the optimal value of maxColors in the DCS scenario.

Number of Readers	Optimal *maxColors*		Throughput in Additive Model		Through-Put Loss
	Single Model	Additive Model	Single Configuration	Additive Configuration	
10	3	4	418.84	474.98	11.8%
20	5	8	96.03	236.71	59.4%
30	7	12	49.10	150.81	67.4%
40	10	17	40.85	109.12	62.6%
50	12	22	28.68	85.15	66.3%

6.4. Percentage of Additive Collisions in the DCS Scenario

It can be observed from Figure 11 that the influence of additive collisions on the performance initially increases up to a peak and then decreases, with respect to the growing value of maxColors . For example with 40 readers, the percentage of additive collisions firstly grows to the peak point at 41% (with maxColors=10), because the direct collisions deceases as maxColors grows, which is also reflected in Figure 10.

Afterwards, the percentage of additive collisions drops down until approaching 0%. In fact, with many time slots, DCS achieves a stable state in which each reader can perform the interrogation without interferences from other readers, since each reader owns a unique time slot.

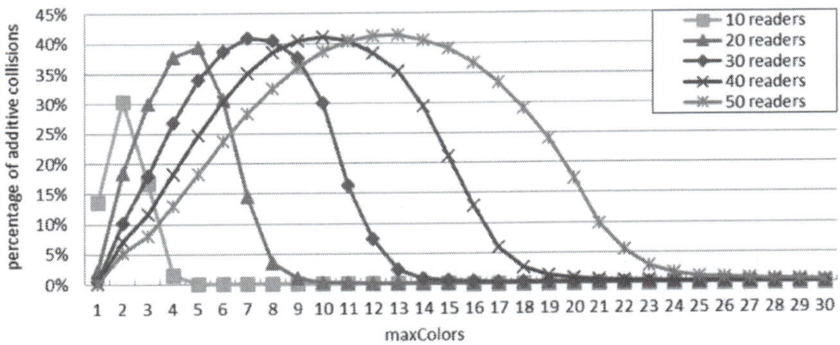

Figure 11. Percentage of additive collisions for the additive interference model in the DCS scenario.

7. CONCLUSIONS

This paper investigates the interference models for the reader-to-reader collisions in RFID systems. In order to evaluate the effects of the models on the estimation of the performance of RFID systems, two general cases, namely the probabilistic interrogations and the slotted interrogations, are considered and simulated. The success rate, the throughput and the percentage of additive collisions are taken into account as analysis metrics. Moreover, DCS is implemented and simulated in order to study the influence of the interference models on a state-of-the-art TDMA anti-collision protocol. The goal of the investigation was to find if and when the simpler single interference models can be used instead of the more accurate additive interference models, without compromising the results of the analysis. In general, a single interference model underestimates the impact of the reader interference on the system performance. Furthermore, in the design of an anti-collision protocol, if the configuration parameters are chosen according to a single interference model, the system may not be able to achieve the optimal performance, as supposed by the additive interference model. Therefore, an accurate study on the RFID reader-to-reader collisions should prefer the additive interference model.

ACKNOWLEDGMENTS

This work has been partially supported by the grant "Nano-materials and -technologies for intelligent monitoring of safety, quality and traceability in confectionery products (NAMATECH)" from Regione Piemonte.

AUTHOR CONTRIBUTIONS

L.Z. developed the simulation tool, performed the experiments and contributed to write the paper. R.F. and F.G. designed the experiments, analyzed the data and contributed to writing the paper. M.R. supervised the research and critically revised the paper.

REFERENCES

1. Hou, J.L.; Chen, T.G. An RFID-based shopping service system for retailers. Adv. Eng. Inform. 2011, 25, 103–115.
2. Piramuthu, S.; Wochner, S.; Grunow, M. Should retail stores also RFID-tag "cheap" items? Eur. J. Oper. Res. 2014, 233, 281–291.
3. Costa, C.; Antonucci, F.; Pallottino, F.; Aguzzi, J.; Sarria, D.; Menesatti, P. A Review on Agri-Food Supply Chain Traceability by Means of RFID Technology. Food Bioprocess Technol. 2013, 6, 353–366.
4. Cuinas, I.; Exposito, I.; Gay-Fernandez, J.A.; Alejos, A.V.; Sanchez, M.G. From Farm to Fork: Traceability Based on RFID—A Proposal

for Complete Traceability in the Wine Sector; SciTePress: Setubal, Portugal, 2012; pp. 265–270.

5. Cimino, M.G.; Marcelloni, F. Autonomic tracing of production processes with mobile and agent-based computing. Inf. Sci. 2011, 181, 935–953.
6. Valero, E.; Adán, A.; Cerrada, C. Evolution of RFID applications in construction: A literature review. Sensors 2015, 15, 15988–16008.
7. Najera, P.; Lopez, J.; Roman, R. Real-time location and inpatient care systems based on passive RFID. J. Netw. Comput. Appl. 2011, 34, 980–989.
8. Badia-Melis, R.; Ruiz-Garcia, L.; Garcia-Hierro, J.; Villalba, J.I.R. Refrigerated fruit storage monitoring combining two different wireless sensing technologies: RFID and WSN. Sensors 2015, 15, 4781–4795.
9. Bueno-Delgado, M.; Vales-Alonso, J.; Angerer, C.; Rupp, M. A comparative study of RFID schedulers in dense reader environments. In Proceedings of the IEEE International Conference on Industrial Technology, Vina del Mar, Chile, 14–17 March 2010; pp. 1373–1378.
10. Papapostolou, A.; Chaouchi, H. RFID-assisted indoor localization and the impact of interference on its performance. J. Netw. Comput. Appl. 2011, 34, 902–913.
11. Bueno-Delgado, M.; Vales-Alonso, J. On the optimal frame-length configuration on real passive RFID systems. J. Netw. Comput. Appl. 2011, 34, 864–876.
12. Shin, K.C.; Park, S.B.; Jo, G.S. Enhanced TDMA based anti-collision algorithm with a dynamic frame size adjustment strategy for mobile RFID readers. Sensors 2009, 9, 845–858.
13. Waldrop, J.; Engels, D.; Sarma, S. Colorwave: A MAC for RFID reader networks. In Proceedings of the IEEE Wireless Communications and Networking, New Orleans, LA, USA, 20 March 2003; Volume 3, pp. 1701–1704.
14. Gilbert, E.N. Random graphs. Ann. Math. Stat. 1959, 30, 1141–1144.
15. Erdős, P.; Rényi, A. On random graphs I. Publ. Math. 1959, 6, 290–297.
16. McDiarmid, C. Random channel assignment in the plane. Random Struct. Algorithms 2003, 22, 187–212.

17. Haenggi, M.; Andrews, J.G.; Baccelli, F.; Dousse, O.; Franceschetti, M. Stochastic geometry and random graphs for the analysis and design of wireless networks. IEEE J. Sel. Areas Commun. 2009, 27, 1029–1046.
18. Díaz, J.; Petit, J.; Serna, M. A random graph model for optical networks of sensors. IEEE Trans. Mob. Comput. 2003, 2, 186–196.
19. Zhang, L.; Ferrero, R.; Gandino, F.; Rebaudengo, M. Evaluation of single and additive interference models for RFID collisions. Math. Comput. Model. 2013, 58, 1236–1248.
20. Gupta, P.; Kumar, P. The capacity of wireless networks. IEEE Trans. Inf. Theory 2000, 46, 388–404.
21. Engels, D.; Sarma, S. The reader collision problem. In Proceedings of the IEEE International Conference on Systems, Man and Cybernetics, Yasmine Hammamet, Tunisia, 6–9 October 2002; Volume 3, pp. 1–6.
22. Kuhn, F.; Wattenhofer, R.; Zollinger, A. Ad-hoc networks beyond unit disk graphs. In Proceedings of the Joint Workshop on Foundations of Mobile Computing, San Diego, CA, USA, 19 September 2003; ACM: New York, NY, USA, 2003; pp. 69–78.
23. Aly, S.A.; Kapoor, V.; Meng, J.; Klappenecker, A. Bounds on the network coding capacity for wireless random networks. In Proceedings of the IEEE Information Theory and Applications Workshop, La Jolla, CA, USA, 29 January–2 February 2007; pp. 231–236.
24. Díaz, J.; Petit, J.; Serna, M. Faulty random geometric networks. Parallel Process. Lett. 2000, 10, 343–357.
25. Díaz, J.; Lotker, Z.; Serna, M. The distant-2 chromatic number of random proximity and random geometric graphs. Inf. Process. Lett. 2008, 106, 144–148.
26. Paredes, R.; Chávez, E. Using the k-nearest neighbor graph for proximity searching in metric spaces. In String Processing and Information Retrieval; Lecture Notes in Computer Science; Springer: Berlin, Germany; Heidelberg, Germany, 2005; Volume 3772, pp. 127–138.
27. Ferrero, R.; Gandino, F.; Montrucchio, B.; Rebaudengo, M.; Zhang, L. A novel simulator for RFID reader-to-reader anti-collision protocols. In Proceedings of the 2015 International EURASIP

Workshop on RFID Technology (EURFID), Rosenheim, Germany, 22–23 October 2015; pp. 1–6.

28. Zhang, L.; Ferrero, R.; Gandino, F.; Rebaudengo, M. Simulation and evaluation of the interference models for RFID reader-to-reader collisions. In Proceedings of the International Conference on Advances in Mobile Computing & Multimedia, Vienna, Austria, 2–4 December 2013; ACM: New York, NY, USA, 2013; pp. 209–216.

CHAPTER 5

RFID-Based Equipment Monitoring System

Mohd Helmy Abd Wahab[1], Herdawatie Abdul Kadir[1], Zarina Tukiran[1], Nor'aisah Sudin[1], Mohd Hafiz A. Jalil[1] and Ayob Johari[1]

[1] Universiti Tun Hussein Onn Malaysia,, Malaysia

1. INTRODUCTION

Automated monitoring systems are becoming trends, creating easier method to identify item, tracking, monitoring and add on security values. In places where there are lots of items accessed by many users, the tendency of loss is high due to weakness in items monitoring. Here, we briefly describe our research on the university's laboratory perspectives. The main aim of the research is to work out a generic approach of monitoring items in a place with several rooms. For example, there are laboratories with expensive equipments are available in a university to support teaching and learning session. Conventional approach of checking items for every session is difficult for lab administrator as most libraries are being used by more than 20 students per session.These leads to a challenge for lab administrator to monitor the flow of these items are always in place. Currently, the monitoring of laboratory equipments is performed manually by the lab administrator during each laboratory sessions. For every loan of equipment, a log book needs to be filled up in order to keep track the transaction information.This system was found to have a lot of weaknesses such as misuse of the equipment log records, losses of equipment, no in-out transaction record and misplace of equipments. To automate the process, Radio Frequency Identification (RFID) is identified as one of the most practical and applicable in real time implementation in-line with the nature where most of the systems are made computerized. In this paper, a solution has been provided for the problem encountered in laboratory equipment monitoring system using RFID technology. Therefore RFID-based monitoring system has been designed and developed to solve the problem associated with the handling of laboratory equipments. This chapter is organised as follows. Section 2 describes related works on RFID-based monitoring system. The architecture of the system is mentioned in section 3. Application scenario and the implementation are briefly explained in section 4 and 5 respectively. Finally, the chapter is concluded in Chapter 6.

2. RELATED WORK ON RFID IN MONITORING

RFID is a wireless automatic identification that is gaining attention and is considered by some to emerge as one of the pervasive computing technologies in history (Roberts, 2006). As the technology grows very rapidly, RFID has received considerable worldwide attention and widely used in monitoring and tracking ranging from human identification to product identification. Previous research has successfully indicated that RFID has been increasingly expanded in various fields such as retail supply chain, asset tracking, postal and courier services, education, construction industry, medical, and etc.

The work presented by Tan and Chang (2010) who had developed an RFID-based e-restaurant system to change the traditional restaurant services which is considered as passive. The utilization of RFID is to improve the service quality which is customer-centered that enable waiters to immediately identify customers via their own RFID-based membership card. It can also provide customized services such as enhanced dining table service; pay the bills, instant transmission of customer orders to kitchens and flexibility of managing payments of bills and discounts. However, in Ngai et. al. (2008), designed and developed RFID-based sushi management system to help a conveyor belt sushi restaurant to achieve better inventory control, responsive replenishment, and food safety control, as well as to improve its quality of service.

In the perspective of animal tracking or livestock monitoring management system, Vouldimos et. al. (2010) developed FARMA project which combined with RFID technology and mobile wireless networking to track animal and the data in repository which contains animal data records. The purposes of the system are to identify animal in case it gets lost and identify some basic information about particular animals. A similar work done by Nor Suryani Bakeri et. al. (2007) and Ahmad Rafiq Adenan et. al. (2006) developed a livestock monitoring system using RFID. An RFID tag is used and attached to each livestock to monitor its movement in and out as well as the basic information about any particular animals.

The use of RFID also could assist in customs clearance process by reducing the delay time. According to Hsu, Shih and Wang (2009), the use of RFID can improve the efficiency of cargo process, and reduce the inventory and labor cost. The work presented based on the mathematical model of the customs clearance process-delay and the network of customs delay is reconstructed based on the use of RFID. RFID also has been successfully applied in global postal and courier services in monitoring the parcel delivery. One of the well known courier service company is DHL which has been using RFID in their services since 1988 and carried out 20 trials on active and passive technology and successfully proved it improved the service and reduce the costs (EPC Global, 2005). The application of RFID in global market in postal and courier services contribute 650 billion per year and Europe was the leader in utilizing RFID in postal and courier services (Zhang, et. al., 2006).

High quality service lead to customer satisfaction, increase market share, and enhance profitability of service organizations (Hoffman and Bateson, 1997). Oztaysi, Baysan, and Akpinar (2009) have done a study to investigate the possibility of using RFID as a tool for improving service quality in hospitality industry and primarily concern in tourism industry.

In monitoring of asset tracking, an effective and efficient managing the tracking of medical-assets in healthcare facilities can be performed by the means of RFID. Oztekin et. al. (2010) has done a study using enhanced maximal covering location problem along with critical index analysis metric to optimize the design of a medical-asset tracking system constrained by a limited number of RFID readers. Results indicate that the proposed technique has improved by 72% compared to the currently utilized expert placement strategy.

Yan and Lee (2009) developed RFID application in Cold Chain monitoring system to track the cold-chain product flowing in supply chain, ensure the products' quality and comply with relevant provisions during transportation. The system executes in real-time environment and can track the location and monitor the temperature of cold-chain products to ensure the quality. However, according to Loebbecke (2005) has done a research regarding the application of RFID in retail supply chain at a brink-and-mortar supermarket to investigate the advantages and challenges with the early RFID applications in terms of technological issue such as standardization, challenges on the data, network and application layers.

Haron, et. al. (2010), designed and developed of a context aware notification system for university students using RFID. The system aims to deliver urgent notifications to the intended students immediately at their respective locations. A quite similar work done by Herdawatie et. al. (2010) which integrates RFID and biometric sensor to track students in a boarding school of their location at the selected restricted area.

As summary, based on the successful of RFID applications in various fields as discussed above, it shows that its application is endless. This section onwards explains the RFID application in tracking of laboratory equipments movement to ensure its availability. It also aims at helping the lab administrator in monitoring the equipment from lost or misplaced. The monitoring of equipments movement is not only being monitored by the lab administrator but also by the top management through online databases.

3. SYSTEM ARCHITECTURE

Building an automated tracking applications by integrating web services guarantee many benefits, such as reduce clerical task and ease the management burden. The RFID-based Equipment Tracking System is an integrated system that offers an effective solution of managing items especially for large scale environment. It combines the RFID technology and security devices to ensure the items are always been monitored and secured. The system enable the university to give admission to selected individual to access locations, permit movement of items, record the important data and also enable the viewing of record via internet.

A faculty usually has a number of laboratories. Faculties with technical courses such as Information Technology and Engineering usually have more laboratories. To implement the system, an appropriate design is required to make sure it is suitable for the number of laboratories and equipments in all laboratories. In this study, the design of the system which utilizes RFID is divided into two; hardware design and software design as shown in the architecture diagram in Figure 1.

There are six important components involved as illustrated in Fig. 1, (1) RFID Tag, (2) RFID Reader, (3) Personal Computer, (4) RS232 Cable, and (5) LAN HUB and 6) CCTV Camera The master server contains the database which is used to store all data collected from RFID reader where user can read or change information in the database. The RFID tags contain antennas to enable the receiving and transferring data. The passive RFID tag creates power from magnetic field and use it to energize the circuits of the RFID chip and sends information back to the reader in the form of radio-frequency waves. The physical layer of the system is depicted in Figure 2. It shows how the computers and the master server are connected. The software involved in developing the system is also outlined.

In the system, RFID technology were implemented to enable data to be automatically recorded where each tag is embedded in the metric card (working pass) for individual and attached to each equipments. The lab administrator will grant an access to selected individual to enter a laboratory and also enable selected individual to move items out from the lab and within the organization. Upon the individual is found attempt to force the process the camera is triggered and activated to document the image of intended person and buzz the alarm system and notify the person-in-charge.

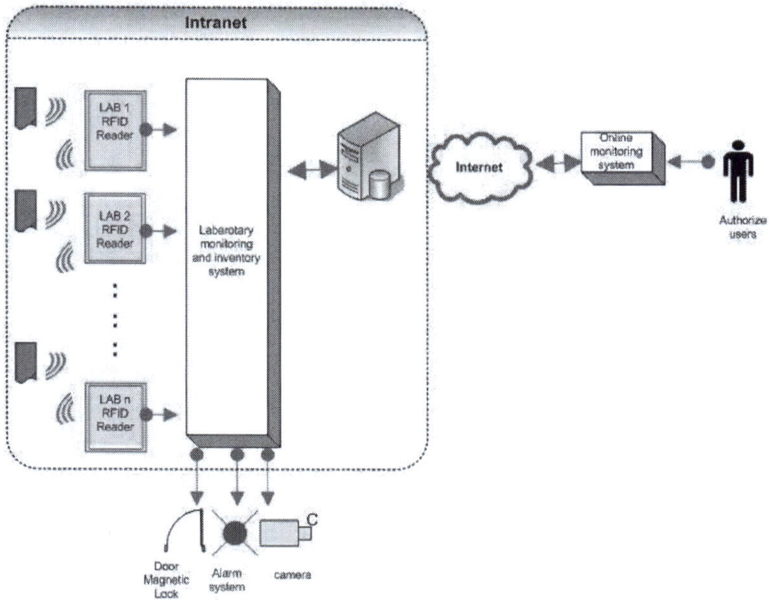

Figure 1. System Architecture

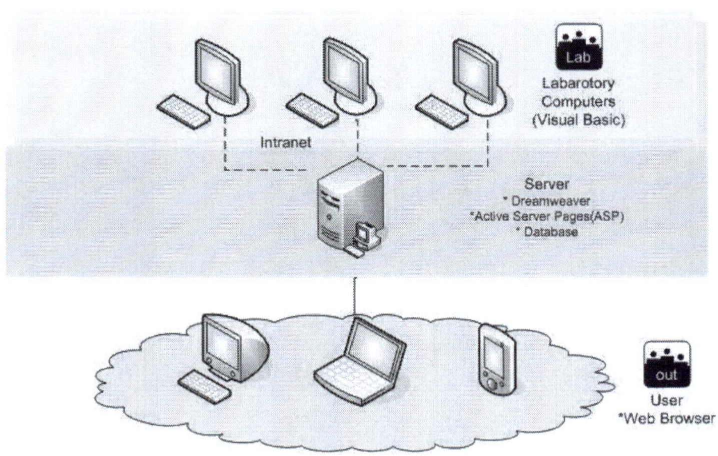

Figure 2. Physical layer of the system

4. APPLICATION SCENARIOS

The developed prototype is an online laboratory monitoring system that has three purposes; which typically composed of (1) Laboratory grant access (2) Inventory control, and (3) Online data viewing. The prototype has been applied at the UTHM research project lab. To illustrate the concept, a sample of layout of application was provided in Figure 3.

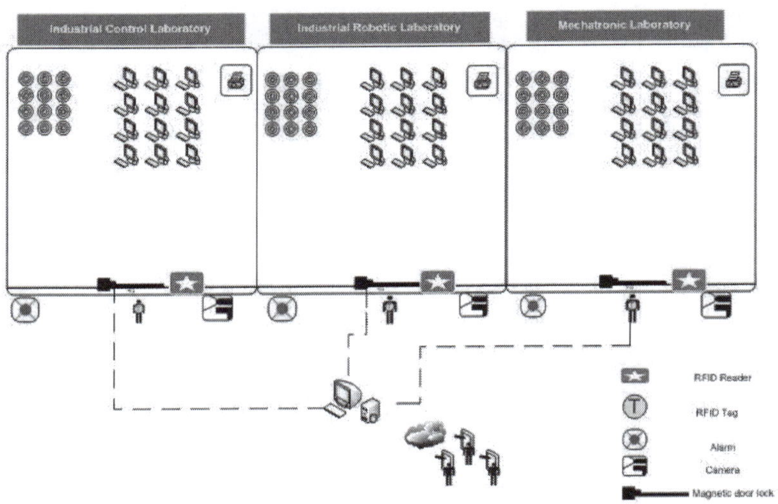

Figure 3. Application scenario layout

In the system, RFID tag is attached to both users and equipments. The RFID reader is located at each Laboratory to record and verify the RFID tags in the area. Each laboratory is equipped with a surveillance camera and an alarm indicator to deal with unforeseen circumstance events. The recorded data is stored and managed by a central computer whereby each laboratory computer is connected via intranet connection to ease any information received from computer lab can be easily transmitted to central computer. The main purpose of data, which is stored at the central computer, is to ease the management to have a look the whereabouts of equipment and record of in-out information. The administrator will grant the personal level access, equipment status and also permit online monitoring to authorize individual.

Legally attempt to enter a laboratory with authorize RFID identification (id), lead the magnetic door to unlock (door open) and record the entry information. Illegally attempt to access the laboratory, the door keep locked and activate the camera and warning sign is indicated to the system.Once the system detects a forceful behavior such as shaking the door, the system triggers an alarm to notify the security. In side of inventory control, intended user must be registered with authorized id before granted to move or lend the equipments, once verified, magnetic door will unlock and information is recorded. Otherwise, the registered id requires re-verification.

5. IMPLEMENTATION

As mentioned in previous section, the RFID-based Equipment Monitoring System is used to keep track the record on laboratory equipment. Hence, the laboratory, its equipments and users who use the equipment in the laboratory need to be part of the system entities. This can be done by enrolling these entities in the system.

At this moment, only three (3) laboratories have used the system and only authorised personnel are allowed to access the labs. In order to ensure only the authorised user login to the labs, they need to present their RFID card. Each laboratory is equipped with magnetic door. Therefore, the RFID card acts as a key to unlock the magnetic door. All equipments placed in the labs are tagged with RFID and registered in the system. This equipment can be used and borrowed by the user either in the same laboratory or in other laboratories. For the latter the user's RFID card and the equipment's tag should be readable by the RFID reader. Once the information is successfully matched by the system, the magnetic door will be unlocked. Otherwise, the door remain locked if only the reader able to read equipment's information but not the user's information. Figure 4 illustrates the flow to access laboratories and equipments.

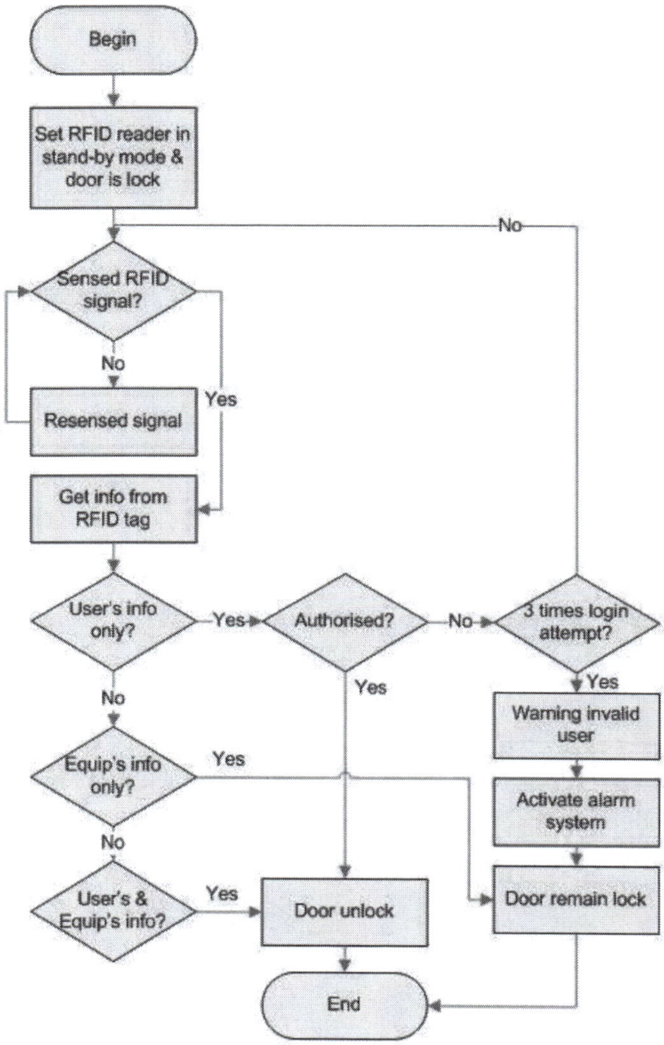

Figure 4. Access laboratories and equipments flowchart

The system has two main purposes; first is to register user, equipment and laboratories to be part of the system entities. This is done by the system administrator through data management module. The second is to keep track the equipment and to monitor the activities of the user. The latter can be accessed through monitoring module. These two main purposes are presented in the form of graphical user interface as shown in Figure 5.

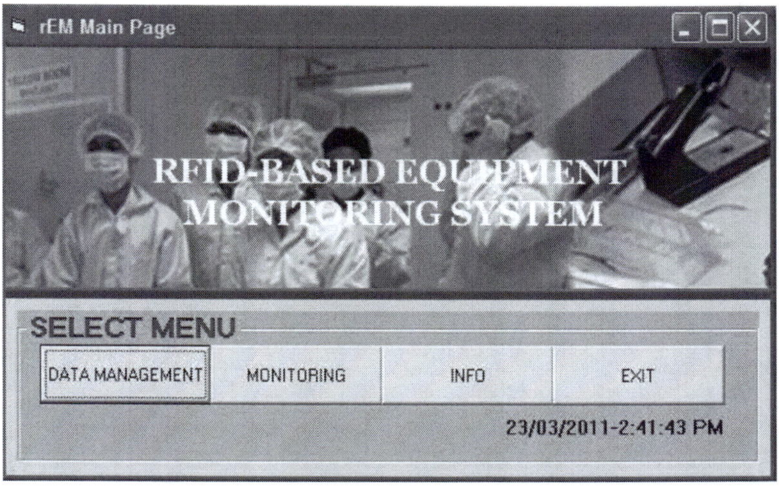

Figure 5. The main GUI of RFID-based Equipment Tracking System

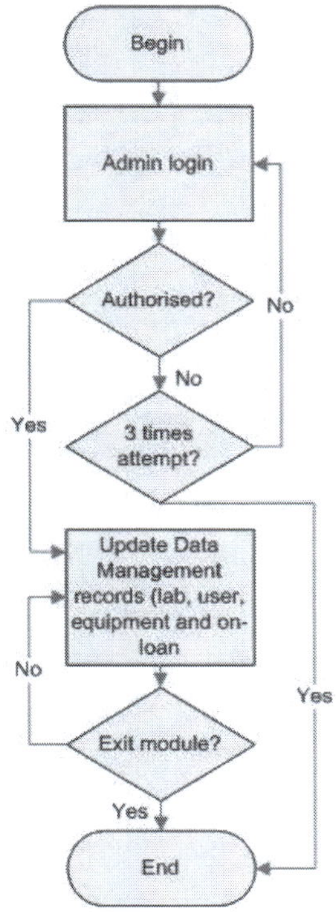

Figure 6. The system flow of data management module

The data management module system flow is illustrated in figure 6. This module can only be accessed by authorised personnel to maintain the integrity of the data. Thus, system administrator needs to enter the correct password in login page as shown in Figure 7. Users are allowed to re-enter the password up to three (3) times for invalid password before the system activates the alarm system.

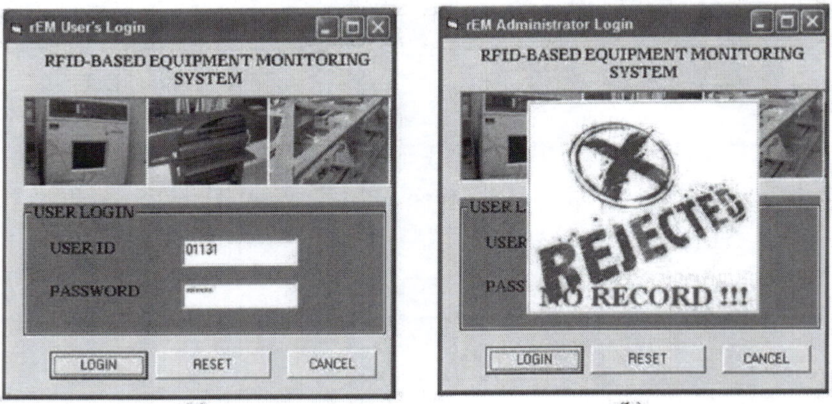

(a) (b)

Figure 7. System administrator's (a) entering the password and (b) warning message for invalid access

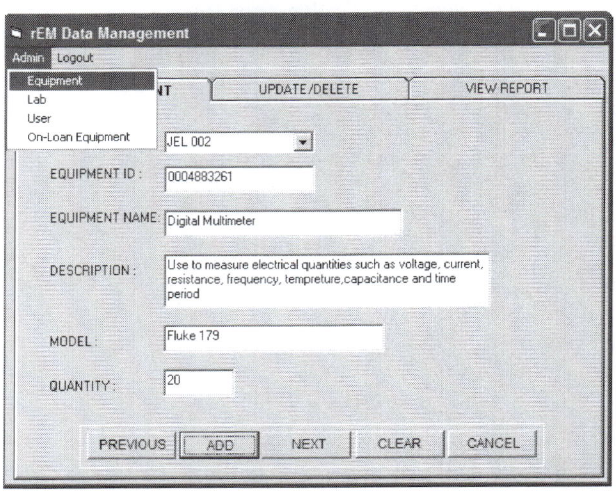

Figure 8. Data Management Module

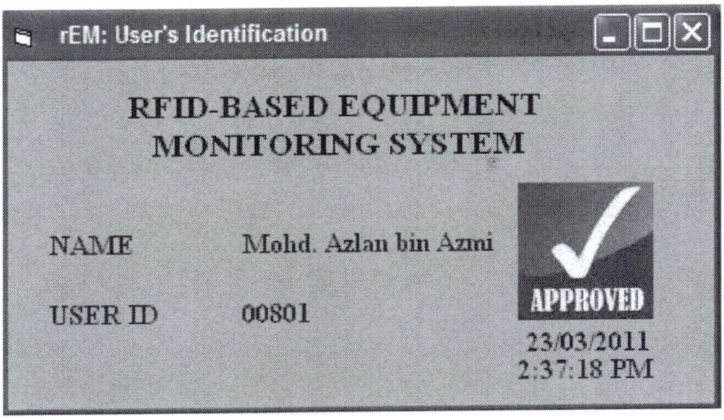

Figure 9. Authorised user

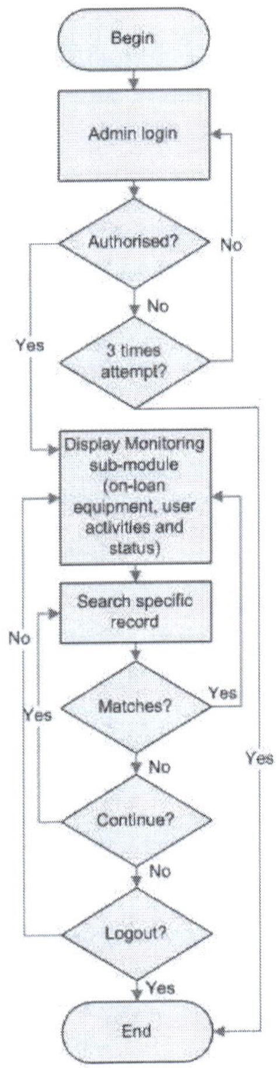

Figure 10. The system flow of data monitoring module

Then system administrator is able to view the data if the administrator is the authorized personnel. The data management module handles four (4) sub-modules which are described as Figure 8. According to Figure 8, *Equipment* sub module allows the system administrator to add new equipment and maintain the equipment record that is assigned in the laboratory. RFID tag is attached on the equipment to track down its status. *Lab* sub module allows the system administrator to enroll any laboratory to the system. *User* sub module allows the system administrator to enrol any user that wants to be in the system. After the user has been registered in the system, each will be given an RFID card. This card is used to authorise access the intended laboratory. The user needs to bring along the card to enter or to leave the laboratory. If user brings in/out any equipment to/from its registered laboratory, the card and the equipment's tag should be read by the RFID reader without fail in order to unlock the magnetic door. Figure 9 shows an example of successful login to laboratory.

On-loan equipment sub module allows the system administrator to register the status of equipment; whether it is in place or it is circulated around laboratory under an authorised user.

The system flow of monitoring module is shown in Figure 10. The module allows the administrator to monitor on-loan equipment, users' activity and their status. This module is designed so that it can be viewed by the administrator internally (intranet access) or remotely (internet access). Here, the discussion is focused on remote access. In order to use this module either internally or remotely, the administrator needs to log-in to the system as shown in Figure 11.

For successful login, the administrator is allowed to view and find a specific record on on-loan equipment. Figure 12 shows on-loan equipment based on laboratory and specific date. As shown below, the following on-loan information is taken from instrumentation lab for Jan 5, 2011. The system is also designed so that the administrator could click on *Equip ID to* view equipment details borrowed by the user.

Figure 13 shows that the administrator is able to view user's activity at each laboratory. In the following example, it shows who has used the instrumentation lab on Jan 5, 2011. The user status tab contains information on which laboratory is allowed and the valid period as shown in Figure 14. By default, this page displays the status of all users. It also could display the status of certain user by selecting specific information, for instance *UserID* keyword to perform the searching process.

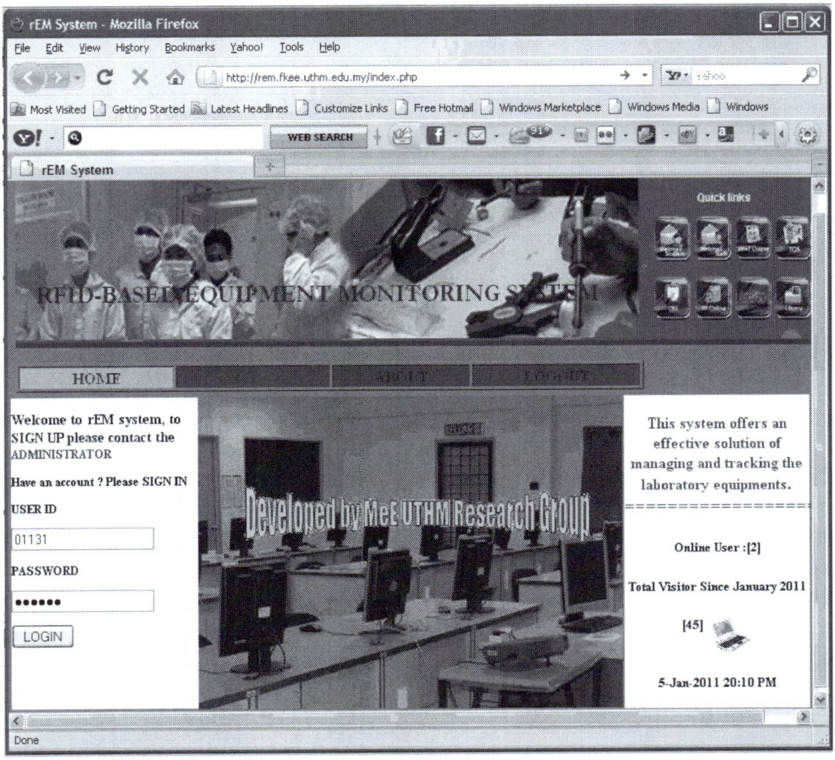

Figure 11. The login page of system

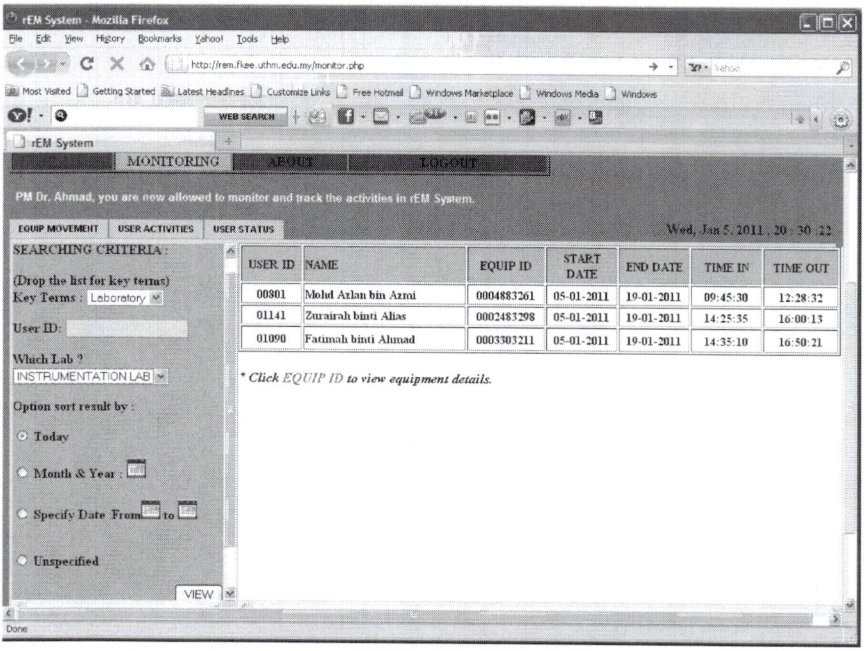

Figure 12. On-loan equipment page

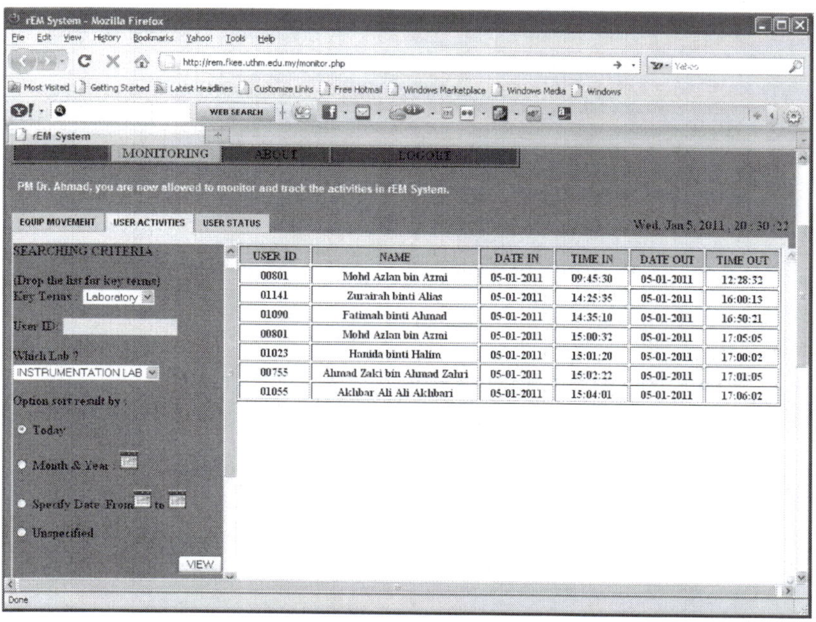

Figure 13. Monitoring user's activity remotely

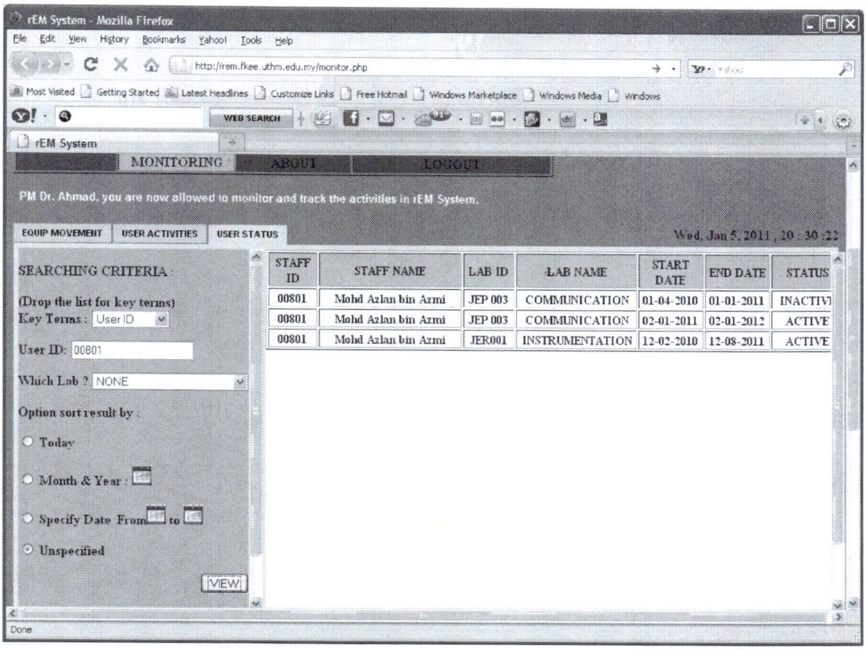

Figure 14. Viewing user's status

6. CONCLUSION

Laboratory equipment monitoring system using RFID is proposed to effectively monitor the in-out equipment from the laboratory. Via this system, every activity involving laboratory equipment can be monitored and updated through web based environment. For security purpose, only authorized personnel have the permit to monitor the transaction activities of laboratory equipment in real-time. The adaptation of RFID-based Equipment Monitoring System also would promote diversity on laboratory management which previously are handled manually.

REFERENCES

1. Ahmad Rafiq Adenan, Siti Zarina Mohd Muji, Mohd Helmy Abd Wahab.Automated Animal Tracking System using Radio Frequency Identification Tags. Proceeding of Computer Science and Mathematics Symposium 2006KUSTEM, Kuala Terengganu, Terengganu, Malaysia, 89November 2006

2. EPC Global. RFID smart label practice experience. 2005 http://www.rfidi--nfo.com. cn/report/dissertation/2OoS08/1655.html

3. N. S. Haron, N. S. Saleem, M. H. Hassan, M. M. Ariffin, I. A. A. R. I. Aziz, Campus. D-based, Notification. Context-Aware, System, Journal of Computing. 23

4. Herdawatie Abdul Kadir, Mohd Helmy Abd Wahab, Zarina Tukiran Mohd Razali Mohd Tomari and Mohd Norzali Hj. 2010, Fusion of Radio Frequency Identification (RFID) and Fingerprint in Boarding School Monitoring System (BoSs), Sustainable Radio Frequency Identification Solutions, Cristina Turcu (Ed.), 978-9-53761-974-9InTech

5. C. Hsu-I, H. Shih-H, W. Wang-C, 2009Applying RFID to Reduce Delay in Import Cargo Customs Clearance Process. Computers & Industrial Engineering. 57506519

6. C. Loebbecke, 2005RFID Technology and Applications in Retail Supply Chain: The Early Metro Group Pilot. 18th Bled eConference on eIntegration in Action, June 6- 8, 2005, Bled, Slovenia.

7. E. W. T. Ngai, S. Y. Y. Lo, 2008Development of an RFID-based sushi management system: The case of a conveyor-belt sushi restaurant. International Journal of Production Economics, 1122630645

8. Nor Suryani Bakery, Ayob Johari, Mohd Helmy Abd Wahab, Danial, Md.Nor. RFID Application in Farming Management System. In Proceeding of 3rd International Conference on Robotics, Vision, Information and Signal Processing 2007ROVISP2007), Penang, 2830November 2007

9. A. Oztekin, F. M. Pajouh, D. Delen, L. K. Swim, 2010An RFID Network Design Methodology for Asset Tracking in Healthcare. Decision Support Systems. 49100109

10. B. Oztaysi, S. Baysan, F. Akpinar, 2009Radio Frequency Identify (RFID) in hospitality. Technovation. 29618624

11. C. M. Robert, 2006Radio Frequency Identification (RFID). Computers & Security, 251826

12. T. Tan-H, C. Chang-S, 2010Development and Evaluation of an RFID-based e-Restaurant System for Customer Centric Service. Expert System with Applications. 379

13. A. S. Voulodimos, C. Z. Patrizakis, A. B. Sideridis, V. A. Ntafis, E. M. Xylouri, 2010A Complete Farm Management System based on Animal Identification using RFID Technology. Computers and Electronics in Agriculture. 70380388

14. B. Yan, D. Lee, 2009Application of RFID in Cold Chain Temperature Monitoring System. 2009 ISECS International Colloqium on Computing, Communication, Control, and Management. Aug. 89Sanya, China.

15. X. Zhang, S. Yue, W. Wang, 2006The Review of RFID Applications in Global Postal and Courier. The Journal of China Universities of Post and Telecommunications. 134

CHAPTER 6

A UHF/UWB hybrid silicon RFID tag with on–chip antennas

Philipp K Gentner[1*], Robert Langwieser[1], Arpad L Scholtz[1], Günter Hofer[2] and Christoph F Mecklenbräuker[1]

1 Institute of Telecommunications, Vienna University of Technology

2 Infineon Technologies Austria AG, Contactless and RF Exploration

ABSTRACT

In this contribution, we describe and analyse a miniature wireless radio frequency identification (RFID) chip with on-chip antennas (OCA) and ultra wideband (UWB) signalling by real-world measurements. With the on-chip antenna approach, no external antennas are required, and the size of the overall tag is identical to the size of the chip alone (3.5 mm×1 mm). The chip is powered through inductive coupling and controlled by an RFID signal at 866 MHz in the downlink, while the uplink transmits a quaternary pulse-position-modulated (4-PPM) UWB signal at 5.64 GHz with pulses having a duration in the order of nanoseconds. In this contribution, the hybrid or asymmetric communication scheme between prototype chip and reader, the embedded OCA, and the measurement setup are described. The prototype achieves 4-PPM bit rate of 126 Mbit/s based on a pulse-train transmission with a duration of 10 μs. The small size, high data rate, and fine time resolution of the UWB impulse radio offer new features and sensing capabilities for future RFID-like applications.

1. INTRODUCTION

Tiny systems on chip (SoC) with on-chip antennas (OCAs) are useful if a transmission of data is necessary over a short distance. Especially for radio frequency identification (RFID) systems, the limited communication range increases security and protects privacy.

Small tags combined with sensing capabilities, such as temperature or shock sensors, allow the tracking of individual items for example frozen goods in a cold chain. Due to their small size, the systems on chip can be placed on or in objects while reducing the cost by avoiding bulky antennas.

In this contribution, a fully integrated system on chip is presented, which features on-chip antennas for a hybrid or in other words asymmetric communication scheme. It gives insight into our past achievements and current studies in the field of short range connectivity devices for RFID applications or for wireless sensors. At first, the scheme is explained in detail in section "Asymmetric communication scheme" with a survey of impulse transmitters and related achievements in research. Then our manufactured ultra-high frequency/ultra-wideband (UHF/UWB) hybrid system on chip is introduced, with all its components and a focus on the on-chip antennas used, a dual antenna concept with an electric and a magnetic antenna.

After the description of the device under test, the development of the measurement setup is explained in detail, and also, results are presented in the time and the frequency domains. The decoding of the received wideband data signal from the hybrid UHF/UWB tag is shown. Before the conclusion, the transmission loss between the impulse-transmitting OCA and the according uplink reader antenna is calculated from measurements.

2. ASYMMETRIC COMMUNICATION SCHEME

The most famous example of an asymmetric communication scheme is used everyday all over the world through browsing of the web. In the downlink from the provider to the user, a much higher amount of data is needed by pictures, movies, or text. In the uplink, mostly control or request data are sent from a personal computer to the provider. This idea is also used in long-term evolution (LTE), where the downlink peak rates are higher as those in the uplink.

In the case of RFID or sensor applications, the high data rate is shifted to the uplink. Since the tags or sensors are very small and the available power has to be used as efficiently as possible, it is advantageous to have a fast transmission in the uplink. The downlink, from a reader station to a sensor or tag, provides the energy, control commands, and clock for synchronisation in a UHF band, with the data rate being usually lower. A concept of an asymmetric communication scheme for RFID scenarios was introduced by Zheng et al. in [1] and has been manufactured and characterised in the works of Baghaei-Nejad and Radiom et al. in [2] and [3].

In the papers mentioned above, impulse radio UWB is considered as the appropriate scheme for the uplink, being able to provide a fast data rate and also being more power efficient. This has been shown by Calhoun et. al in [4], who summarise the necessary energy per pulse for UWB transmitters of recent research. The authors found that the energy per pulse is lower than 10nJ/bit and constant over a data rate from 1 Mbit/s to 10 Gbit/s, and compares it with the energy per bit for a bluetooth transmitter being typically 25 nJ/bit.

In our opinion, transmitting pulses in the baseband is not a good choice and should not be considered for the application in mind. The reasons are the following:

- The powering UHF field interferes with the spectrum of pulses transmitted in the baseband, which leads to an increased reader complexity.

- Reading multiple tags at the same time leads to collisions. Using modulated pulses, the probability of collisions can be reduced by changing the modulation frequency.
- A small size reader antenna is important for many applications.
- Using on-chip antennas with their small aperture, uplink frequencies in the C- or X-band are preferred.

Non-baseband impulse radio UWB transmitters can be realised in two ways [5]: firstly, baseband pulses can be up-converted with the aid of a local oscillator. Secondly, pulse creation, e.g. with delay lines following pulse shaping filter can be employed. In terms of reduced transmitter power consumption, a pulse-position modulation scheme is very promising because it allows to transmit multiple bits per pulse.

Impulse radio as an additional asset of a tag introduces a multitude of new possibilities and applications for RFID scenarios [6]. In recent UWB research, indoor localisation is studied extensively for example by Meissner et al. in [7]. Concerning localisation, the IEEE standard 802.15.4a provides a good basis for commercial applications. Zheng et al. describe in [8] a transceiver achieving a ranging accuracy of 3 cm. Readers for localisation will very likely use an array antenna to track the small device. In [9] and [10], UWB antenna arrays are introduced and analysed concerning their ability to operate in a wideband spectrum and are consecutively used to operate with the wideband tags.

Two inductively coupled magnetic loop antennas offer a wireless power transfer technique, which is used successfully for medical implants [11], such as pacemakers and biosensors, or for charging electric vehicles [12]. An RFID tag with an asymmetric communication scheme favourably should store energy from an energy source to be able to transmit pulses. Here, power-scavenging units harvest energy from the narrowband UHF downlink signal. Figures of merits are the efficiency of the rectifier and the value of the capacitor, where the harvested energy is stored. The downlink antenna will usually be an inductive antenna with high Q-factor, whereas a wideband antenna with low Q-factor is required in the uplink.

3. HYBRID UHF/UWB SYSTEM ON CHIP

In our approach towards a short-range connectivity device, we started with externally powered complementary metal oxide semiconductor substrate (CMOS) structures (see Figure 1). These CMOS prototypes with OCAs were controlled and powered through bondwires. In order to compare the radiation of different types of OCAs, a voltage-controlled oscillator (VCO) was swept over a huge frequency range of several octaves [13]. Furthermore, nanosecond pulses were transmitted with a simple on-off keying pulse transmitter - created by a glitch generator, whose output was multiplied with the VCO frequency. A proof of concept concerning the energy transfer was done, with inductive coupling using a magnetic OCA [14]. With a modulated backscatter technique, the voltage induced into the miniaturised loop antenna was analysed over area and over distance. The power harvesting concept was studied and characterised with a small 3 mm × 1 mm system, where two different OCAs were used for receiving energy and for transmitting data [15]. The power-harvesting concept must be able to collect power of some μW in the long term. This energy, stored in an efficient way with on-chip batteries or capacitors, is used for driving the UWB front end, which requires some mW within a short period of time. The asymmetric scheme allows to replace the used harvesting concept, if clock and data for configuring the chip is provided through another communication channel. Also, this scheme allows to investigate the performance of on-chip antennas without having a galvanic connection. In the course of these investigations, we observed that bondwires used in earlier work have a huge influence due to coupling. Bondwires, typically the same size and in close vicinity of the chip, act as a high Q-factor antenna, and support either the transmitted pulses or the energy-harvesting path. In the downlink path, a coupling into the chips' V_{SS} is established, which yields to a better efficiency of the rectifier. Using plane OCAs, the energy received by the reader (uplink path) was much lower than in the early experiments employing bondwires.

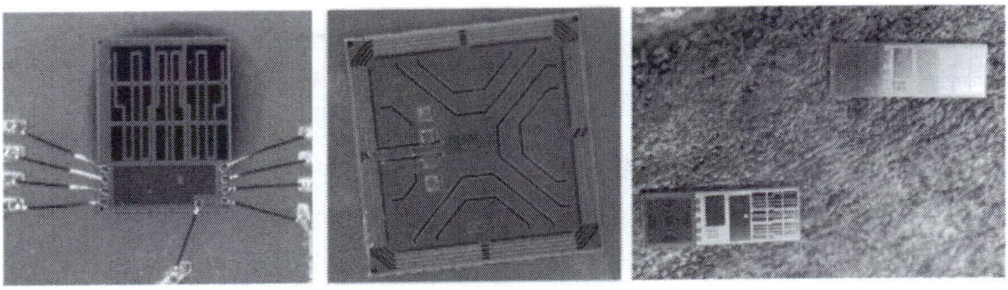

Figure 1. Microscopic images of on-chip antennas (OCAs) From left to right: external powered PPM UWB modulator front end with a meandered dipole on-chip antenna [13]; loop antenna on chip for inductive coupling [14]; prototype of a passive tag with two OCAs [15].

4. FUNCTIONALITY OF THE SYSTEM ON CHIP

The device under test shown in detail in this contribution is a full system on chip. A transmitter, a receiver, a power harvester, and antennas are integrated within one host platform, which is placed on one piece of silicon. The device used for short-range connectivity were manufactured with a standard 130 nm CMOS technology. A loop structure and an electric monopole were integrated on the chip as can be seen in the chip photograph and antenna schematic in Figure 2. The loop structure is connected to a harvester and extracts the necessary energy to drive the circuit from a UHF signal by inductive coupling. The monopole, placed inside the loop antenna, is a meandered line which transmits the UWB pulses. With respect to the overall size of the SoC (3.5 mm × 1 mm), the antenna part is small (1 mm^2). The circuit on chip is programmable via the UHF link, and the signal sent on the uplink can be selected to be a continuous wave, a data sequence of pulses, or a continuous pulse stream. For characterising the device under test, the flexibility to be able to select different waveforms is helpful to detect the low-power transmitted signal. Compared to other work with OCAs and UWB transmitters [16] in the field, this SoC harvests and stores energy as a system and is operated and controlled without any wired contact [2].

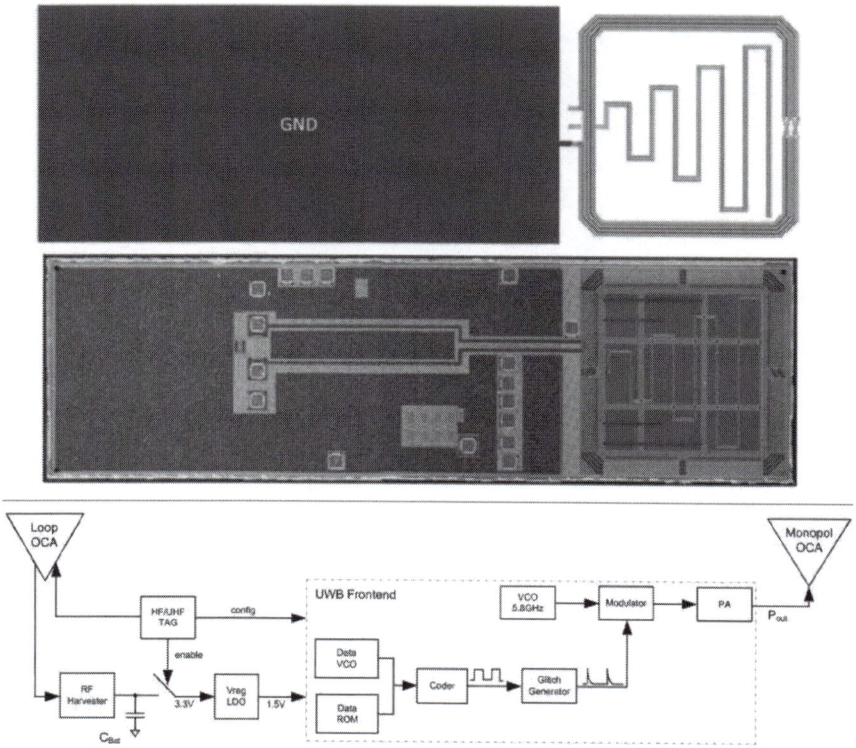

Figure 2. The silicon under test is a UHF/UWB hybrid system on chip with on-chip antennas. From the top to the bottom, a drawing of the antennas, a microscope picture of the system on chip, and a functional schematic of the integrated circuit.

5. ON-CHIP ANTENNAS

In the early research of on-chip antennas, these antennas were considered as replacements of wired clock distribution [17] or as wireless interconnects on chip [18, 19]. OCAs are very common in the millimeter-wave region [20] because this is their native environment.

In the chip presented, a meandered monopole is placed inside a loop antenna, and both are located within the chip area. The antennas are placed on the chip in a way that no electrical circuits are below the antenna. The structure is shown in top of Figure 2. To support the monopole, the remaining part of the silicon is covered with metal and connected to the chips' V_{SS}, which acts as the ground plane. A single-ended antenna is used to avoid a differential power amplification in the front end. This reduces the overall power consumption.

The silicon substrate thickness is 220 μm, with the conductivity being larger than 50 S/m. In [21], where exemplary dipoles have been simulated on a CMOS layer stack, it has been shown that the lossy substrate reduces the Q-factor of the antenna and therefore supports a wideband communication scheme, even when the aperture of the antenna is very small.

All antennas are placed in a so-called inductive layer in which metal filling structures between the turns are avoided. The width of the conductor used to form the monopole is 15 μm with a total unwrapped length of 4.85 mm which is approximately $\lambda/10$. The distance between each turn is 90 μm. With increasing distance to the excitation port, the width of the meandered monopole increases from 90 to 800 μm in eight steps. For the uplink and the downlink antenna, three metal layers each are connected in parallel by vias to increase the thickness of the conductor. The magnetic loop antenna with four turns and chamfered edges is square with a width of 989 μm. The conductor used for the loop has a width of 15 μm and a spacing of 2.6 μm. With a method of moments (MOM) simulator (Sonnet Software, North Syracuse, NY, USA) the inductance of the loop was calculated to be 50.2 nH, and the impedance of the meandered monopole was determined as $Z_{\text{uplinkMOM}} = (11.6 - 6.6j)\ \Omega$. The accurateness of the simulation result is dependent on the physical size of the antenna and the simulation method itself, especially for small wires such as the conductors for the OCAs [22]. Therefore, and for further investigations, the meandered monopole was simulated with a finite element method (FEM) solver (HFSS - Ansoft, Ansys, Inc., Canonsburg, PA, USA). The impedance thus obtained is $Z_{\text{uplinkFEM}} = (23.4 + 3.66j)\ \Omega$. The real part of the numerically calculated impedance from the FEM simulation is larger because the metallic antenna parts must be modelled as solid aluminium. This is in contrast to the MOM simulation, where all thin metal layers are accurately modelled as a combination of copper and aluminium. Therefore, the impedance's real part is lower in the MOM simulation. In comparison to theoretical predictions for electrically small antennas, the real part is very much dominated by the losses due to the conductive silicon substrate underneath.

The far field gain of the monopole is determined to be -41.5 dBi with an efficiency of -43.56 dB at the targeted centre frequency of 5.8 GHz. The gain flatness is 0.9 dB from 5 to 6.2 GHz which is important for pulse transmission.

6. INTEGRATED CIRCUIT

In the downlink from the reader to the miniature device, a standard UHF RFID EPC protocol is used. The communication frequency is $f_c = 866$ MHz. A power harvester [23] stores the energy in a buffer capacitor with a capacitance of $C_{Bat} = 3.9$ nF. The HF rectifier of [23] is replaced by a standard UHF rectifier. For the functionality of this component, it is not important if the tag is placed in the near or far field of a reader antenna. Moreover the used electric or magnetic antenna connected to rectifier is important, so that the tag can be operated either in the near or far field of a reader. The harvester needs a minimum input of $P_{min} = -15$ dBm for operation. The DC current consumption of the tag is 8.5 µA. Simulations show that a power of $P = -10$ dBm, not needed by the transponder circuitry, is redirected into the charge pump, placed before the buffer capacitor. A schematic of the tag is shown in the bottom of Figure 2.

Once the capacitor is fully charged and the command to transmit the hardwired sequence is triggered by the RFID reader, pulses with a nominal width of either $w_{low} = 0.78$ ns or $w_{high} = 1.3$ ns are created by the glitch generator and transmitted by the monopole antenna. The uplink frequency is $f_c = 5.8$ GHz. The motivation of this uplink frequency is not triggered by standardisation; furthermore, it is a trade-off between CMOS transit frequency and hence a low power consumption of the active front end. Since a carrier-based pulse radio is considered in the tag, a centre frequency within an available ISM band is feasible. A pulse-position modulation is used, where four sub frames and one guard frame form one symbol frame (4-PPM). The transmitted data sequence consists of 16 symbols, which are repeated until the capacitor reaches a minimum level of charge. In the design, the data rate was calculated to be 117

Mbit/s and the nominal symbol rate is 58 Msym/s. The supply voltage of the UWB front end is 1.5 V, controlled by a low-dropout (LDO) regulator placed before the UWB front end. The available power P_{out} provided by the single-ended amplifier to the OCA monopole was simulated using Cadence software to be -1.3 dBm. The simulated antenna impedance is connected in Cadence to the front end, and the peak output power of a continuous wave delivered by the front end is determined. In the following measurements, the pulse mode is analysed, hence the available power has to be corrected with the duty cycle of the pulse signal. Therefore, in pulse mode, with a pulse width of $w_{low} = 0.78$ ns and a duty cycle of $D = 4.5\%$, the available power fed to the OCA is $P_{out} = -28.2$ dBm.

7. MEASUREMENTS WITH A HYBRID UHF/UWB READER SETUP

A photograph of the measurement setup is shown in Figure 3. A UHF RFID reader (RFID ME™ USB Dongle, MTI, Inc., Hsinchu, Taiwan) acts as the signal source for the downlink path. The USB reader was programmed to be able to send control commands and select the different operating modes of the silicon under test.

Since the pulse burst transmitted by the silicon is as short as 10 μs, a triggering event is required. Therefore, the initial rising edge of the UHF signal is sampled by a coupler and consequently used as trigger source for the oscilloscope. Following the coupler two lowpass filters in series suppress the spurious harmonics produced by the UHF reader.

The scenario is shown in the subfigure in Figure 3: The downlink excitation coil is printed on a standard FR4 substrate and is matched to the downlink path (centre frequency 866 MHz). This external loop antenna replaces the internal antenna of the UHF reader. In this way, we have more degrees of freedom for characterising the device under test. The UWB silicon chip is fixed with adhesive tape in close vicinity to the excitation coil. We wish to point out, however, that the chip has not a single conductive connection to any part of the measurement equipment.

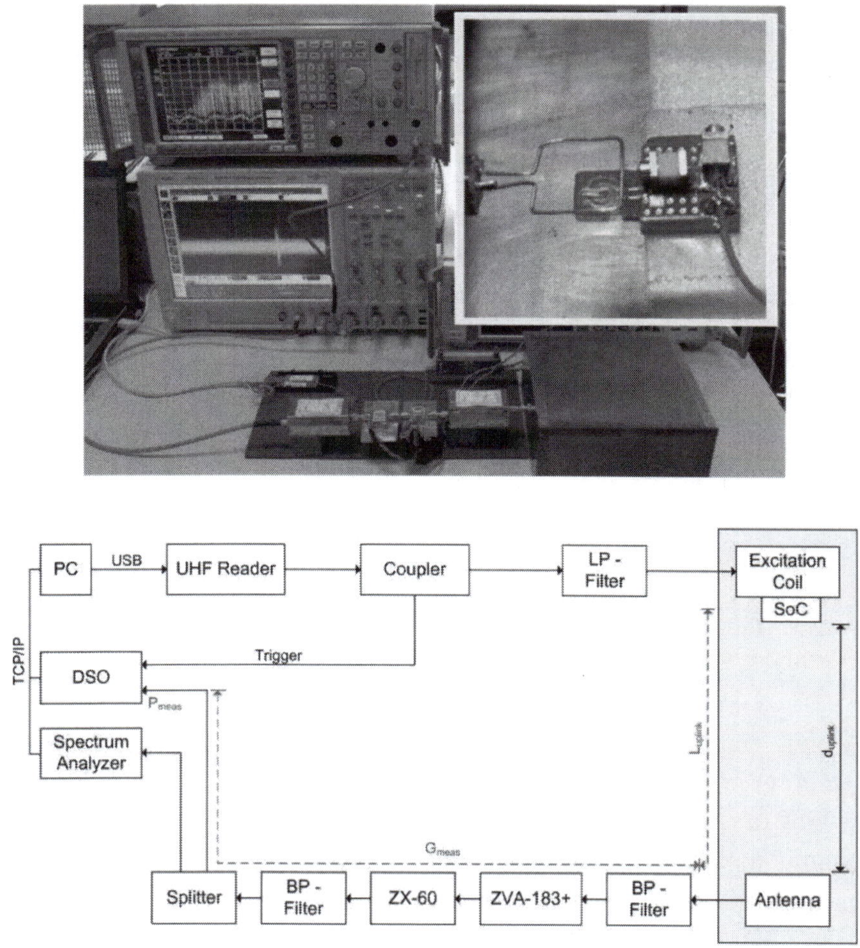

Figure 3. Photograph and schematic of the hybrid RFID/UWB reader measurement setup used for decoding of the chips' transmitted UWB sequence.

A reference antenna was positioned in a distance of $d_{\text{uplink}} \approx 5$ mm above the silicon, capturing the silicon's uplink. This reference antenna is also a loop structure with one turn of copper wire; designed to be resonant at the uplink frequency of 5.8 GHz. The SoC, the UHF excitation coil, and the reference antenna were located inside a brass box to suppress the local WLAN and other interferers.

As shown in schematic view of the measurement setup (Figure 3), a chain of bandpass filters and amplifiers is used before the signal is fed to a digital sampling oscilloscope.

The components of the uplink, such as the reference coil and the two parallel coupled half-wave resonators, are measured and the results presented in Figure 4. All passive components have been simulated and measured with respect to a reference of 50 Ω. The transmission coefficients of the bandpass filters are $G_{BP1} = -1.3$ dB and $G_{BP2} = -1.4$ dB. The influence of the splitter and the cables is summarised in $G_{sc} = -7.5$ dB. The two amplifiers have a gain of $G_{ZX60} = 12.3$ dB and $G_{ZVA183+} = 27.2$ dB. The amplification of the total uplink path is

$$G_{meas} = G_{ZX60} + G_{ZVA183+} + G_{BP1} + G_{BP2} + G_{sc} \quad (1)$$

which results in $G_{meas} = 29.3$ dB. The power at the oscilloscope or spectrum analyser can be written as

$$P_{meas} = G_{meas} + L_{uplink} + P_{out}. \quad (2)$$

L_{uplink} describes the transmission between the monopole and the reader antenna, including the gain and possible mismatch of both antennas.

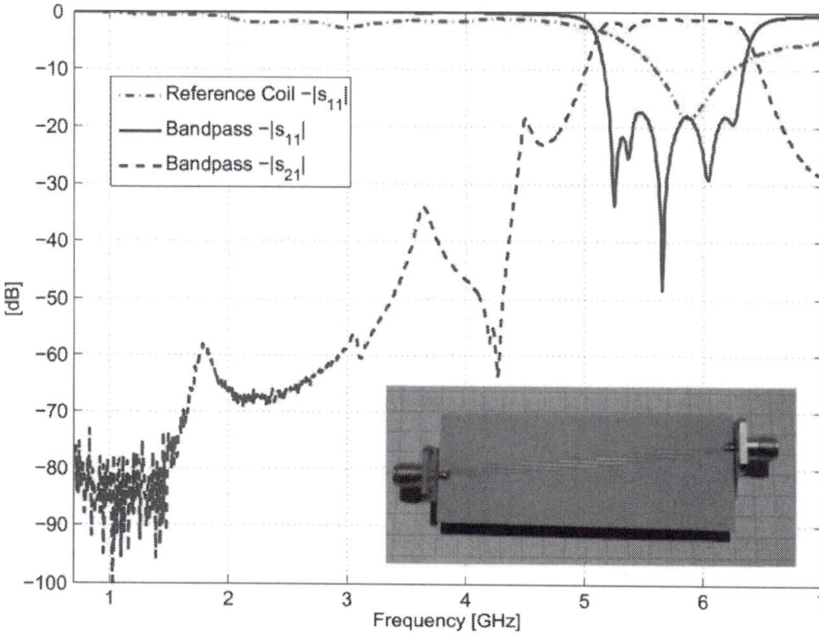

Figure 4. Logarithmic transmission and reflection coefficients of reference coil and bandpass filters. These were used in the uplink path of the measurement setup.

8. FREQUENCY DOMAIN MEASUREMENT

A spectrum analyser was used to characterise the uplink in the frequency domain. A resolution bandwidth of 10 MHz and a video bandwidth of 50 kHz were set. Since the impulse radio event of the tag is short in time and the sweep time of the analyser is too long to capture a transmission precisely, the max hold functionality was set and 32,500 sweeps were captured. The result of this measurement is shown in Figure 5. We observe that the harmonics of the 866 MHz downlink signal are present at 5,196 MHz and 6,062 MHz, indicating that more suppression in the downlink path of the measurement setup is necessary. The tag is set to transmit continuous pulses (constant pulse repetition rate), hence a comb line spectrum is expected. This spectrum is visible at a centre frequency of $f_c = 5.64$ GHz, which differs only slightly from the targeted design frequency due to CMOS process variation. The spectral line at this centre frequency is 13 dB above the noise floor. The 10 dB bandwidth of the pulsed signal is 230 MHz. The coil antenna used for receiving the pulse signal is narrowing the observed frequency band. This can be seen in Figure 4 with the plotted reflection coefficients of the reference coil.

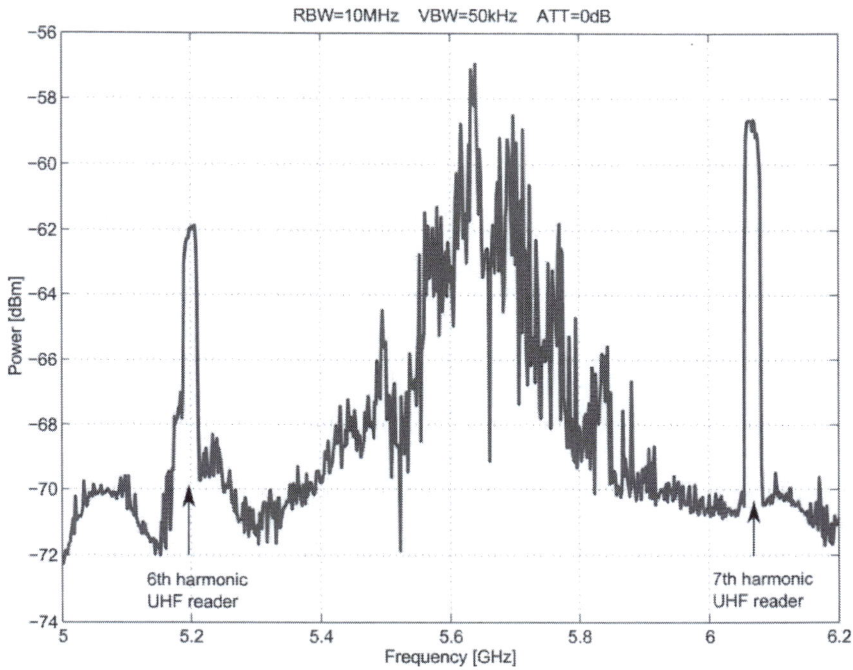

Figure 5. Measurement of wideband uplink signal with harmonics of narrowband downlink signal in frequency domain.

9. TIME DOMAIN MEASUREMENT

Two captures with a digital storage oscilloscope (40 GSa/s) as shown in Figure 6 were taken. In part (a) of Figure 6, a continuous stream of fixed 4-PPM symbols is presented. A 16-symbol data like sequence is shown in part (b). For both, the complete burst is shown on the left side. The right side is a window zoomed into the respective waveforms. The continuous pulse stream has an average frame length of 15.8 ns, which yields to a data rate of 63.3 MSym/s. Comparing the envelope of the burst amplitude of the continuous signal and the data sequence, one can observe that the latter is almost constant in time between 0 and 8 μs, whereas the amplitude of the continuous signal is initially higher and decreases with time. From transient simulations, it is known that at ≈ 8 μs, the initial voltage of 3.3 V stored in the buffer capacitor discharges below the minimal supply voltage of 1.5 V, required by the UWB front end. The average amplitude of the continuous pulsed signal is higher than that of the waveform with the data sequence. The amplitude of the preamble, which is based on five continuous pulses, is higher than the amplitude of the following data sequence.

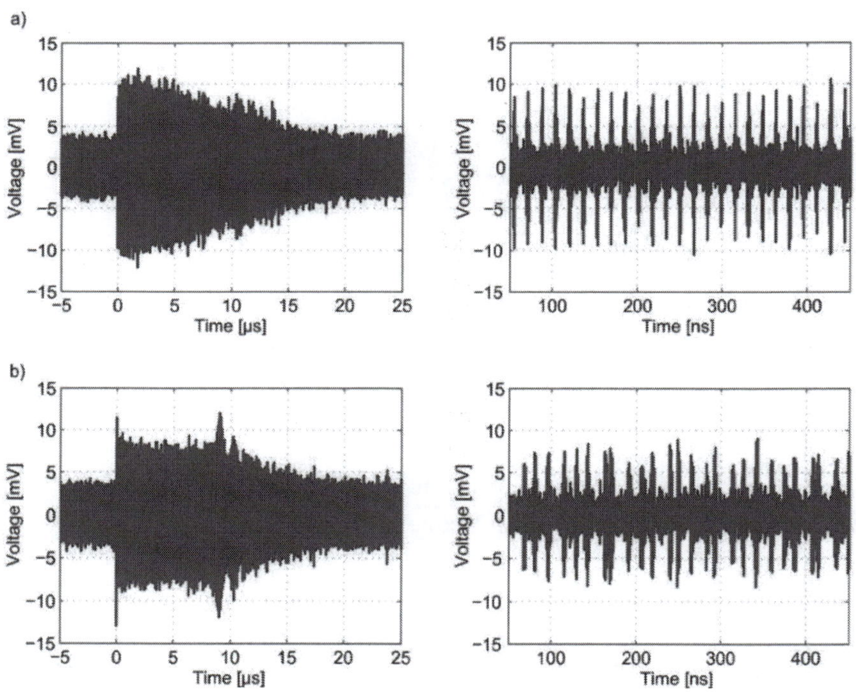

Figure 6. Captured voltage waveforms in the time domain. With amplitude of the miniature wireless chip transmitting a continuous **(a)** and data **(b)** sequence.

10. DECODING OF THE UWB UPLINK 4-PPM SIGNAL

In this section, the 4-PPM signal is decoded step by step in a Matlab post processing step. The oscilloscope captured the waveform of the transmitted pulses, with the pulse width set to w_{low} in the data sequence mode. Not to waste processing time, from the received signal $r(t) = s(t) + n(t)$ the pulse burst is extracted. This burst is squared and filtered with a Matlab FIR equiripple bandpass filter. The filter has a centre frequency of 5.6 GHz and a 60 dB bandwidth of 1.6 GHz. Now, the signal is squared again and subsequently normalised to the maximum peak.

$$r'(t) = \left(r(t)^2 * h_{\mathrm{BPF}}\right)^2 \tag{3}$$

$$x(t) = \frac{1}{\max\left(r'(t)\right)} \times r'(t) \tag{4}$$

The normalised signal $x(t)$ is shown in Figure 7a. In this representation, one can observe a deep fade in the amplitude at around 10 µs. At this instant, the voltage supplied by the on-chip capacitor becomes too weak for the UWB front end to operate.

The decoder itself is based on determining the time between the current pulse and the following pulse, taking into account the estimated symbol corresponding to the current pulse. Since a preamble of five symbols is transmitted at the beginning of the pulse burst, an initial state for decoding is given. In general, the amount of possible pulse distances n_Δ in an M-ary pulse position modulation scheme is given by

$$n_\Delta = 2M - 1. \tag{5}$$

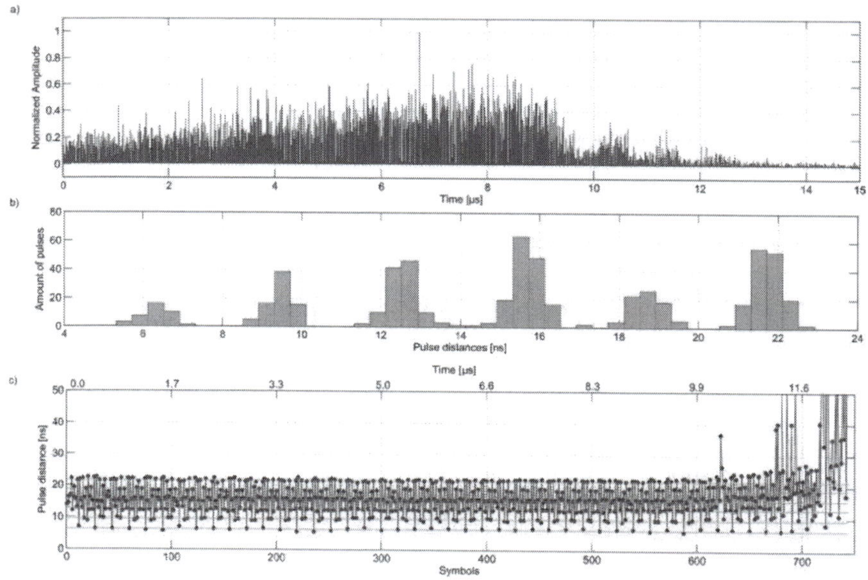

Figure 7 Amplitude of pulse burst, pulse distances of transmitted pulse burst, and pulse distances versus transmitted symbols. Filtered and normalised amplitude of the pulse burst **(a)**, histogram of pulse distances t_Δ of a transmitted pulse burst **(b)**, and the pulse distances versus transmitted symbols **(c)**.

The pulse distances are calculated after a threshold comparator. In Figure 7b the pulse distances t_Δ of the first 600 symbols (≈10 µs) are shown as a histogram. It shows the occurrence of pulse distances within the captured pulse burst. The amount of pulses having a pulse distance of 15.8 ns is maximum across all pulse distances. It represents consecutive equal symbols in the modulation scheme and being the frame length of the 4-PPM transmission. Only $n_\Delta = 6$ distances can be found in the burst. The reason is that one out of seven possible transitions is not present in the hardwired 4-PPM transmitted data. Clear peaks can be observed in this representation. We conclude that no significant timing drift is caused by a decreasing supply voltage to the UWB front end.

All occurring pulse distances t_Δ in the data sequence of the burst are plotted versus symbols together with the average t_Δ in Figure 7c. In this graph, one can observe that not only the frame rate ($t_\Delta = 15.8$ ns) but all other pulse distances are constant until 620 symbols. The pulse distances of all following symbols drift away and a successful decoding is not possible any more.

In the following, the transmitted symbols are estimated, with estimation borders positioned between the average pulse distances. The symbol error rate is calculated for pulse burst transmissions of several consecutive measurements. The result is shown in Figure 8 as symbol error versus decoded symbols. In most cases, a transmission without errors is possible until 600 symbols (equal to 1,200 bits). The theoretical and experimental limitation to decode up to the given number of symbols is given through the finite size of the buffer capacitor of the SoC, and the power efficiency of the TX-UWB front end, being the only consumer.

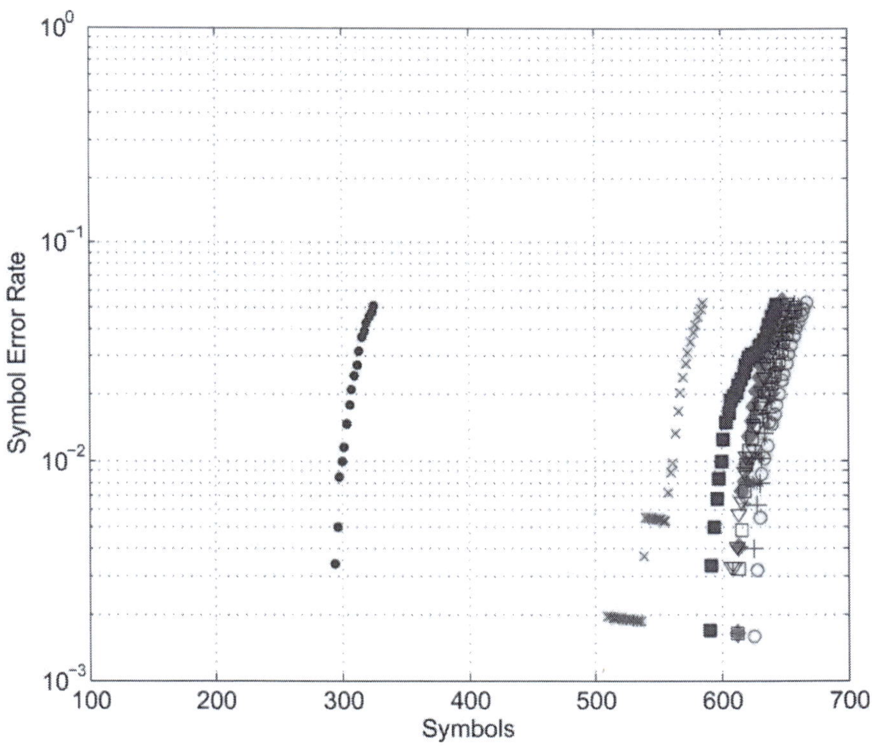

Figure 8. Symbol error rate versus transmitted symbols. For nine consecutive UWB data transmissions from SoC to reader antenna.

11. UWB OCA TRANSMISSION LOSS

The received power P_{meas} can be calculated with the root mean square voltage of the signal and the duty cycle of the impulse signal [24]:

$$U_{RMS} = \frac{U_P}{\sqrt{2}} \times D \tag{6}$$

$$P_{meas} = \frac{U^2_{RMS}}{R} = \frac{1}{2R} \times \left(\frac{U_P \times T}{\tau}\right)^2. \tag{7}$$

With an average pulse amplitude of $U_p = 8$ mV and a duty cycle of $D = T/\tau = 2.6$ ns/15.8 ns $= 0.16$ it is $P_{meas} = -47.6$ dBm. The transmission coefficient between the monopole and the readers' loop antenna according to this measurement is $L_{uplink} = -48.7$ dB.

12. COMPARISON WITH RELATED WORK

The figures of merit of the presented device using a UHF/UWB asymmetric communication scheme are compared to similar devices in Table 1. It has to be mentioned that the device from [16] is powered externally, and the devices in [25] and [6] use external antennas. In this table, the required power $P_{downlink}$ necessary to establish a wideband connectivity over the distance $d_{downlink}$ is used as a basis of comparison.

Table 1. Important figures of merit for RFID tag with hybrid communication scheme in comparison to similar research

Path	Parameter	This work	Radiom [3]	Kulkarni [16]	Pelissier [25]	Vauche [6]
	Technology	130 nm	180 nm	180 nm	130 nm	130 nm
	Die - Size	3.5 mm^2	4.5 mm^2	> 5.4 mm^2	4 mm^2	0.8 mm^2
Downlink	OCA	Yes	Yes	-	Ext. antenna	Ext. antenna
	$f_{downlink}$	866 MHz	5.8 GHz	-	900 MHz	900 MHz
	Receiver	Yes	Yes	-	Yes	UHF and UWB
	Antenna gain	-	−29.5 dBi ([2])	-	-	0 dBi
	$P_{downlink}$	13 dBm	36 dBm	-	14 dBm	36 dBm
	$d_{downlink}$	0 mm	75 mm	-	100 mm	10 m
Harvester	Capacitor	3.9 nF	6 nF	-	-	56 pF
Uplink	OCA	Yes	Yes	Yes	Ext. antenna	Ext. antenna
	f_{uplink}	5.64 GHz	5.8 GHz	8 GHz	7.9 GHz	7.25 GHz (sim)
	Antenna gain	−41.5 dBi	-	−27.4 dBi	-	- (sim)
	IR-scheme	4 PPM	OOK / BPSK	BPM	OOK	Localisation (sim)
	Data rate	126.6 MBit/s	1 MBit/s	750 MBit/s	112.5 Mbit/s	200 Mbit/s (sim)
	Pulse width	2.6 ns	1 ns	0.5 ns	2.9 ns	0.45 ns (sim)
	P_{uplink}	−28.2 dBm	-	−42 dBm	−16.6 dBm	-
	d_{uplink}	5 mm	75 mm	-	100 mm	-
	Bits per burst	1,248	-	-	-	1 (sim)

CONCLUSIONS

In this contribution, the potential use of on-chip antennas for RFID tags with ultra-wideband signalling is investigated by a prototypical implementation. The SoC is described in detail featuring two on-chip antennas and an asymmetric wireless communication scheme.

The measurement setup is presented which comprises a commercially available UHF RFID dongle and a spectrum analyzer as well as a digital storage oscilloscope. An optimisation is carried out in the uplink path for increasing the signal-to-noise ratio of the received signal using bandpass filters and amplifiers. Further, interferers in the downlink path (e.g. local WLAN and harmonics of the powering UHF signal) had to be suppressed.

Bursts of nanosecond pulses from the UHF/UWB hybrid silicon RFID tag were captured and analysed with the optimised measurement setup. The time domain and frequency domain measurements were subsequently analysed by post processing. Due to the short transmission time, the low duty cycle and the low transmit power of the signal, we found that it is easier to detect the radiated signal in the time domain. The ability to switch the SoC between different operating modes for data transmission enhances the initial signal detection.

The transmitted ultra-wideband data sequence from the tiny SoC is decoded correctly up to sequence lengths of 1,248 bits. The embedded monopole OCA broaden the pulse width from 0.7 to 2.6 ns. A data transmission with simulated pulse width of 1.3 ns is therefore not successful.

The transmission loss from a CMOS OCA to a reader antenna is estimated to be $L_{uplink} = -48.7$ dB from transient simulations and measurements.

In comparison to related work in the field [3], the downlink distance is reduced which is reflected in a much lower power requirement for inductive coupling by 20 dB. We conclude that Impulse Radio is a promising low-power communication scheme for a fast, secure, and power-efficient data transfer for future RFID enhancements.

ACKNOWLEDGEMENTS

This work was performed as part of the project ‘ConSens’ (Contactless Sensing) within the funding programme ‘Forschung, Innovation, Technologie - Informationstechnologie’ (FIT-IT) of the ‘Bundesministerium für Verkehr, Innovation und Technologie’ (BMVIT). The authors wish to thank the members of the COST action IC 1004, “Cooperative Radio Communications for Green Smart Environments,” for countless fruitful discussions.

REFERENCES

1. Zheng LR, Nejad MB, Zou Z, DS Zou X, Zhang Z, Tenhunen H: **Future RFID and wireless sensors for ubiquitous intelligence.** In Proc NORCHIP. Tallinn; 16-17 Nov. 2008:142-149.

2. Baghaei-Nejad M, Radiom S, Vandenbosch G, Zheng LR, Gielen G: **Fully integrated 1.2 pJ/p UWB transmitter with on-chip antenna for wireless identification.** In IEEE International Conference on Ultra-Wideband. Nanjing; 20-23 Sept. 2010:1-4.

3. Radiom S, Baghaei-Nejad M, Aghdam K, Vandenbosch G, Zheng L, Gielen G: **Far-field on-chip antennas monolithically integrated in a wireless-powered 5.8-GHz downlink/UWB uplink RFID tag in 0.18 µm standard CMOS.** IEEE J. Solid-State Circ 2010,**45**(9):1746-1758.View Article

4. Calhoun BH, Lach J, Stankovic J, Wentzloff DD, Whitehouse K, Barth AT, Brown JK, Li Q, Oh S, Roberts NE, Zhang Y: **Body sensor networks: a holistic approach from silicon to users.** In Proceedings of the IEEE 2012, **100:** 91-106.View Article

5. Wentzloff D, Chandrakasan A: **Gaussian pulse generators for subbanded ultra-wideband transmitters.** IEEE Trans. Microwave Theory and Tech 2006,**54**(4):1647-1655.View Article

6. Vauche R, Bergeret E, Gaubert J, Bourdel S, Fourquin O, Dehaese N: **A remotely UHF powered UWB transmitter for high precision localization of RFID tag.** In IEEE International Conference on Ultra-Wideband. Bologna; 14–16 Sept 2011:494-498.

7. Meissner P, Arnitz D, Gigl T, Witrisal K: **Analysis of an indoor UWB channel for multipath-aided localization.** In IEEE International Conference on Ultra-Wideband. Bologna; 14–16 Sept. 2011:565-569.

8. Zheng Y, Arasu MA, Wong KW, The YJ, Poh AHS, Tran DD, Yeoh WG, Kwong DL: **A 0.18µm CMOS 802.15.4a UWB Transceiver for Communication and Localization.** In Solid-State Circuits Conference. San Francisco; 3–7 Feb 2008:118-120.

9. Gentner PK, Hilton GS, Beach MA, Mecklenbräuker CF: **Characterisation of ultra-wideband antenna arrays with spacings following a geometric progression.** IET Commun 2012,**6**(10):1179-1186. 10.1049/iet-com.2010.1054MATHMathSciNetView Article

10. Gvozdenovic NM, Gentner PK, Mecklenbräuker CF: **Antenna array for the reader of an ultra-wideband identification tag with on-chip antenna.** In Loughborough Antennas & Propagation Conference. Loughborough; 14–15 Nov. 2011:1-4.

11. Mahfouz M, To G, Kuhn M: **No strings attached.** IEEE Microwave Mag., 2011,**12**(7):S34-S48.View Article

12. Shinohara N: **Power without wires.** IEEE Microwave Mag. 2011,**12**(7):S64-S73.MathSciNetView Article

13. Gentner PK, Wiessflecker M, Arthaber H, Scholtz AL, Mecklenbräuker CF: **Measured wideband near-field characteristics of an UWB RFID tag with on-chip antenna.** In International Conference on Ultra-Wideband. Bologna; 14–16 Sept. 2011:479-483.

14. Gentner PK, Hofer G, Scholtz AL, Mecklenbräuker CF: **Accurate measurement of power transfer to an RFID tag with on-chip antenna.** In Progress In Electromagnetics Research Symp. Moscow; 18–21 Aug. 2012:227-230.

15. Gentner PK, Amreich P, Reinisch H, Hofer G: **A passive ultra wideband tag for radio frequency identification or wireless sensor networks.** In International Conference on Ultra-Wideband. Syracuse; 17–20 Sept. 2012:2-5.

16. Kulkarni V, Muqsith M: **A 750 Mb/s, 12 pJ/b, 6-to-10 GHz CMOS IR-UWB transmitter with embedded on-chip antenna.** IEEE J. Solid-State Circ. 2009,**44**(2):394-403.View Article

17. Kim K, Yoon H, O KK: **On-chip wireless interconnection with integrated antennas.** In Electron Devices Meeting. San Francisco; 10–13 Dec. 2000:485-488.

18. Wang Y, Makadia D, Margala M: **On-chip integrated antennas - the first challenge for reliable on-chip wireless interconnects.** In Canadian Conference on Electrical and Computer Engineering. Ottawa; 7–10 May 2006:2322-2325.

19. Kikkawa T, Saha PK, Sasaki N, Kimoto K: **Gaussian monocycle pulse transmitter using 0.18 μm CMOS technology with on-chip integrated antennas for inter-chip UWB communication.** IEEE J. Solid-State Circ. 2008,**43**(5):1303-1312.View Article

20. Yao T, Tchoketch-Kebir L, Yuryevich O, Voinigescu SP, Gordon M: **65GHz Doppler Sensor with On-Chip Antenna in 0.18μm SiGe BiCMOS.** In MTT-S International Microwave Symposium Digest (2006). San Francisco; 11–16 June 2006:1493-1496.View Article

21. Gentner PK, Adalan A, Scholtz AL, Mecklenbräuker CF: **Impact analysis of silicon and bondwires on an on-chip antenna.** In European Conference on Antennas and Propagation. Prague; 26–30 March 2012:3168-3172.

22. Estarki MD, Xing Y, Vaughan RG, XU H: **The effect of gap size on dipole impedance using the induced EMF method.** URSI International Symposium on Electromagnetic Theory (EMTS) Berlin, 16–19 Aug., (IEEE, Piscataway, 2010), pp. 373–376

23. Reinisch H, Gruber S, Wiessflecker M, Unterassinger H, Hofer G, Pribyl W, Holweg G: **An electro-magnetic energy harvester with 190nW idle mode power consumption for wireless sensor nodes.** In European Solid-State Circ. Conference. Seville; 14–16 Sept. 2010:234-237.

24. Oppermann I, Hämäläinen M, Iinatti J (Eds): UWB: Theory and Applications. Wiley, West Sussex; 2004.

25. Pelissier M, Jantunen J, Gomez B, Arponen J, Masson G, Dia S, Varteva J, Gary M: **A 112 Mb/s full duplex remotely-powered impulse-UWB RFID transceiver for wireless NV-memory applications.** IEEE J. Solid-State Circuits 2011,**46**(4):916-927.

CHAPTER 7

Using RFID Technology for Simplification of Retail Processes

Azra Bayraktar[1], Erdal Yılmaz and Şakir Erdem

[1] Marmara University, School of Economics and Business Administration, Department of Marketing and Production Management, Turkey

1. INTRODUCTION

"*Technology has made our lives more full, yet at the same time we've become uncomfortably "full"*. (Maeda, 2006, p.I). After considering these words within marketing context we can see that each day we are dealing with an information flow. Although there is a lot of information about almost everything (sold products, barcodes, invoices, information from the supplier, prices, customer data, competitors etc.) today's managers are more unsecure to take certain decisions. They also don't have enough time to pay attention for these controllable or uncontrollable forces. During a business process the way from production to wholesalers and than to retailers is very complicated. After products finally meet the customers, feedbacks are coming back to the companies and the cycle begins from the start. This product and information flow makes business processes very complex because different people (engineers, sales staff, consumers, managers) interfere this cycle. In this study our aim is - based on the simplicity theory of John Maeda – with the help of RFID technology to create a simple process model for retailers. By using RFID tags in their warehouses and stores they might be able to serve better and more efficiently to their customers and have a better overview in a short period of time. The information supplied via RFID allows corporations to plan their internal processes more efficiently. We also would like to analyze the pitfalls of RFID with a case from Turkish retail industry especially for In-Store usage of RFID.

2. RFID IMPLICATIONS AND SIMPLIFICATION OF PROCESSES

In his theory Maeda emphasizes simplicity laws: (Maeda, 2006, p.I)

Reduce: The simplest way to achieve simplicity is through thoughtful reduction

Organize: Organization makes a system of many appear fewer

Time: Savings in time feel like simplicity

Learn: Knowledge makes everything simpler

Differences: Simplicity and complexity need each other ….

1. Reducing: Reducing the unnecessary information through RFID tacked products and shelves.
2. Organize: Organizing the sales person more effective so they can have necessary information about their customers with the help of wireless handheld devices and can read the personalized customer cards.
3. Time: With the help of RFID tagged products the store managers can gather information about stock levels and responds to customer requirements on time.

4. Learn: Store managers and marketing managers can learn detailed and up to minute product information during the day.

5. Differences: Not every customer is the same; marketers should approach different customers in different ways.

6. Shrinking the time: With the help of RFID- you can get crucial and accurate information about your customer preferences on time.

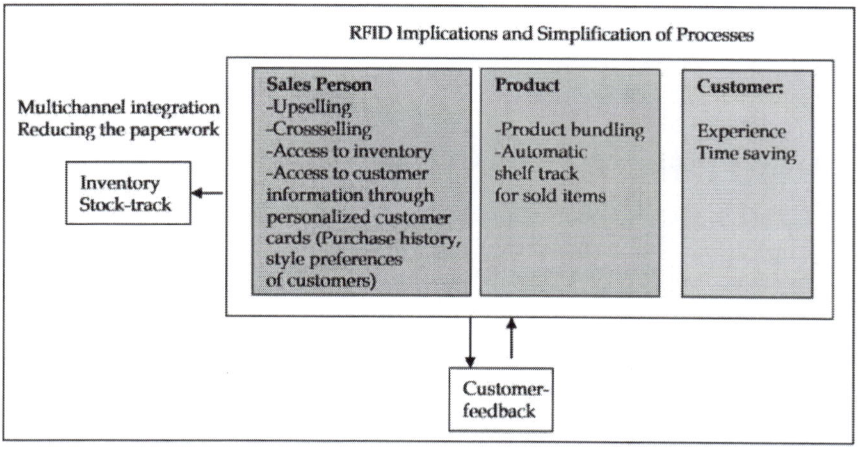

Figure 1. RFID Implications and Simplification of Processes

2.1. The definition of RFID

RFID is a form of automatic identification and data capture (AIDC) technology that uses electric or magnetic fields at radio frequencies for identification, authentication, location, or automatic data acquisition and transmit, and support a wide range of applications—everything from asset management and tracking to access control and automated payment. RFID systems have the capability of sharing information across organizational boundaries, such as supply chain applications (Sabbaghi and Vaidyanathan, 2008, p.73).

Radio Frequency Identification (RFID) is a generic technology concept that refers to the use of radio waves to identify objects (Auto-ID Center

2002). The core of RFID technology is the RFID transponder (tag) – a tiny computer chip with an antenna. Consumer good suppliers attach these tags to logistic units (palettes, cases, cartons and hanger-good shipments) and, in some cases, to individual items. Logistic units and individual items are identified by the Electronic Product Code (EPC). An RFID reader is used to identify the EPC stored on the RFID tag. The antenna enables the microchip to transmit the object information to the reader, which transforms it to a format understandable by computers (Angelles, 2005, p. 52).

Empowered by the capability to identify uniquely and automatically provide continuous, accurate and real time information on the position and the status of product instances, RFID offers a great improvement opportunity to the shelf replenishment process (Bardaki, Pramateri, 2008; p:4)

2.2. Components of RFID systems

RFID Technologies support a wide range of applications—everything from asset management and tracking to manufactured products and related customer services to access controls and automated payments. Each RFID system has different components and customizations so that it can support a particular business process for an enterprise. Depending on the application in an industry and the enterprise within an industry, A RFID system can be very complex, and its implementations may vary greatly. Conceptually, RFID system may be composed of three subsystems as shown in the figure below (Sabbaghi and Vaidyanathan, 2008, p.73):

1. An RF subsystem, which performs identification and related transactions using wireless communication,

2. An enterprise subsystem, which contains computers running specialized software that can store, process, and analyze data acquired from RF subsystem transactions to make the data useful to a supported business process, and

3. An inter-enterprise subsystem, which connects enterprise subsystems when information needs to be shared across organizational boundaries.

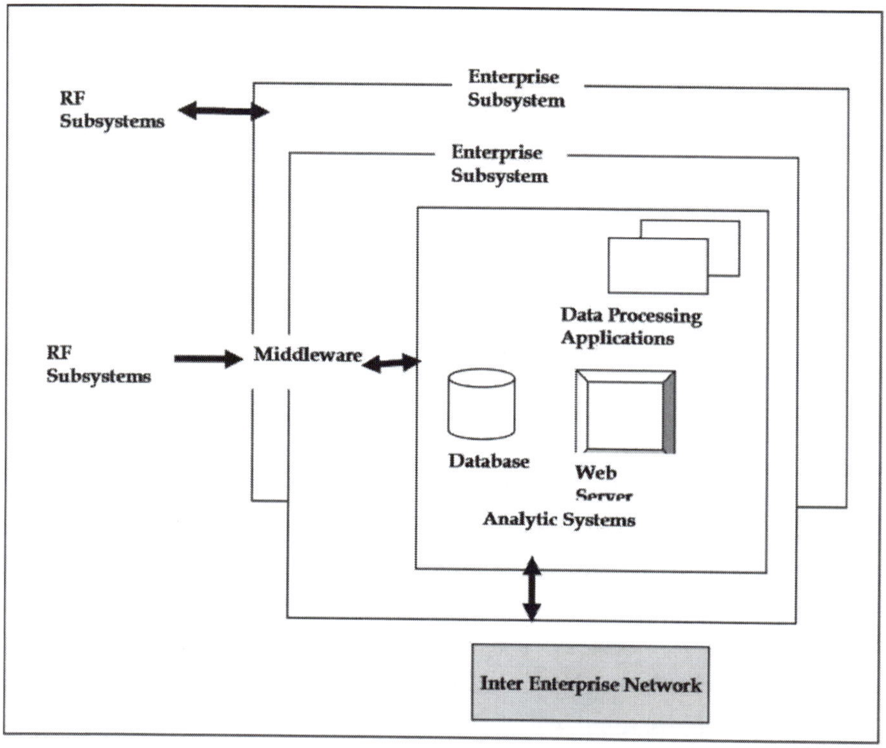

Figure 2. Inter-Enterprise Architecture

Asghar Sabbaghi and Ganesh Vaidyanathan "Effectiveness and Efficiency of RFID technology in Supply Chain Management: Strategic Values and Challenges", Journal of Theoretical and Applied Electronic Commerce Research ISSN 0718–1876 Electronic Version Vol. 3 / Issue 2 / August 2008 / 71-81

Every RFID system contains an RF subsystem, which is composed of tags and readers. In many RFID systems, the RF subsystem is supported by an enterprise subsystem that is composed of middleware, analytic systems, and networking services. However, in a supply chain application, a tagged product is tracked throughout its life cycle, from the manufacture to final purchase, and sometimes even afterwards (e.g., to support targeted product recalls or related service), and thus its RFID systems has to share information across organizational boundaries. Thus, the RFID systems supporting supply chain applications have also an inter-enterprise subsystem (Sabbaghi and Ganesh, 2008, p 71-81).

3. WAREHOUSE APPLICATIONS FOR RFID

RFID can be used for many warehouse inventory management operations, including receiving, storage, picking and shipping procedures. With RFID system, items can have a unique and secure serial numbers and it became so visible in inventory and supply chain operations. This visibility brings several benefits that can eliminate current disadvantages and this wireless system can be more efficient.

Warehousing Operations:

Receiving; when pallets are unloaded from the truck, they are automatically identified with fixed position or mobile RFID readers. Fixed position RFID readers can be mounted at the dock door. Mobile readers can be designed as a PDA or they can be mounted on a forklift. Mobile RFID readers can be more effective because they can be used throughout the facility and they require less investment. While RFID system is integrated with Warehouse Management System (WMS), data read from the pallet's tags are transferred into WMS and updating inventory files. This warehouse process reduces the labor needs. If bar codes were being used in this process, all received pallets would have to be scanned by workers and on the contrary RFID, bar codes needs clearly visible labels.

Storage; in the conventional warehouse systems, different items should be storage different locations. Whereas RFID readers can scan locations and read RFID tags from anywhere. As a result of this, items do not have to be storage in specific locations. In this way, many different storage location alternatives can be used for fast replenishment and picking.

Picking; when RFID system integrates with order management system, the order is checked by the WMS to confirm the picked item belongs with the order.

Shipping; the order management system can confirm pallet loads and improve the accuracy of the shipping process with RFID readers. A RFID tagged pallet can be identified a fixed position RFID reader (as vehicle mounted) or a mobile reader (handheld device). RFID allows for an automatic check of the items loaded into the trailer against the customer order. (Jones and Chung, 2008, pg. 325)

Mandates from the large retailers (Wal-Mart, Target, Albertson etc) and government agencies in USA have increased the awareness of RFID. But still companies that have not been affected by the mandate requirements prefer to wait until the technology matures so that they have adequate knowledge about its potential benefits, especially many companies are concerned about the ROI (Return on Investment) models of RFID (Bhattacharya et al. 2007, p. 1; Jabjiniak and Gilbert, 2004). Determining accurate measures for RFID ROI is very important in order to convince managers. Developing a comprehensive framework for all short term and long term benefits will contribute toward the development of ROI measures (Bhattacharya, 2007, p.2).

4. BENEFITS OF RFID SYSTEM IN RETAILING

RFID technology can track inventory more accurately in real time resulting in reduced processing time and labor. There are many applications and possibilities for RFID/EPC as these objects in motion are traced throughout the supply chain. The complete visibility of accurate inventory data throughout the supply chain from manufacturer's shop floor to warehouses to retail stores brings opportunities for improvement and transformation in various processes of the supply chain. RFID technology can help a wide range of organizations and individuals such as hospitals and patients, retailers and customers, and manufacturers and distributors throughout the supply chain to realize significant productivity gains and efficiencies (Sabbaghi and Vaidyanathan, 2008, p.72).

RFID usage in retailing has taken a lot of attention recently (Bhattacharya et al.2007) because the retail industry is one of the most aggressive supporters of this technology; In comparison, a 2005 report by Frost & Sullivan determined the revenue in the RFID retail market to be $400.2 million in 2004, a figure expected to grow to $4,169 million by 2011 (Bacheldor, 2006, p.1).

During the last decade several research studies have focused on RFID and its benefits and challenges in retail sector (Bhattacharya, 2007, p.3) :

Table 1. Studies about RFID benefits in Retail Sector

Larsson and Qviberg (2004)[1]	Justification of RFID implementation
Jones et al. (2004)[2]	Potential benefits and challenges of RFID throughout the supply chain for retailers in UK.
Koh, Kim and Kim (2006)[3]	Issues and critical factors of RFID in retail industry
Vijayaraman and Osyk (2006) [4]	Empirical study of RFID implementation in warehousing industry
Karkkainen (2003) [5]	Analysis of RFID benefits obtained by increasing supply chain efficiency for short shelf life products.

Bhattacharya, Mithu; Chu Chao-Hsien; Mullen Tracy (2007). RFID Implementation in Retail Industry: Current Status, Issues and Challenges; *Decision Science Institute (DSI) Conference*, Phoenix Arizona AZ, p.3

According to IdTechEx (Bhattacharya, 2007, p.6) the retail industry will comprise %44 of the global RFID market value system including tags by the year 2016. In retail industry RFID is expected to replace the barcode technology as it provides more benefits. The most important benefits for the future is integrated supply chain management, which enables availability of products, inventory management and decreasing of the costs (Bhattacharya, 2007, p.6). After conducting a content analysis in 2006 about RFID and retrieved 362 articles which have been published between 2002 and 2006 Bhattacharya *et al.* have categorized and summarized the challenges, drivers and benefits of RFID technology for retail industry along with the frequency of articles that support the analysis (Bhattacharya, 2007, p.9- 15)

Table 2. Benefits from RFID Implementation in Retail Industry

Drivers for RFID in Retail Ind.	Percentage
Benefits* Wal-Mart Mandate Decreasing Cost of tags and readers EPC Global initiatives for Standardization Anti-Counter feiting TOTAL	92,68% 2.85% 2.03% 2.03% 0,41% 100%
Benefits* from RFID Implementation	**Percentage**
Operational Efficiency - Reduced out of stock - Accuracy, speed and efficiency of process - Automated shipping/receiving - Reduced Inventory - Improved efficiency of store operations - Improved labor productivity - Streamlined process achievement/Leaner manufacturing	36.84%
Improved Visibility - Real-time Visibility - Tracking and Tracing - Improved visibility of orders and inventory - Asset Management - Return/Recall Management - Tracking shopping behavior - Streamlined reverse logistics	24.12%
Reduced Costs - Reduced labor requirements/costs - Reduced overall costs	10.53%
Improved Security Security against theft/fraud Reduced shrinkage Improved supply chain security Eliminates return merchandise fraud	9.21%
Improved Customer Service Levels	7.89%
Better Information Accuracy Improved packing and shipment accuracy Business Intelligence	7.46%
Increased Sales	3.95%

Bhattacharya, Mithu; Chu Chao-Hsien; Mullen Tracy (2007). RFID Implementation in Retail Industry: Current Status, Issues and Challenges; *Decision Science Institute (DSI) Conference*, Phoenix Arizona AZ, 2007, (pp 9-11).

Bhattacharya *et al.* has also emphasized that all these retailer specific benefits have a potential to impact customer service levels positively. Although RFID imbedded loyalty programs can add extra values for the customers, store managers and shopping mall management, many companies are acting reluctantly still to use RFID for their Customer Relationship Strategies (Bayraktar; Yilmaz, 2010).

Privacy issues are one of the main concerns of RFID. A balance should be kept between the benefits that consumers can get in terms of better service, time saving and protection of their privacy. Also many business managers would like to see a detailed cost and benefit analysis of RFID implementation.

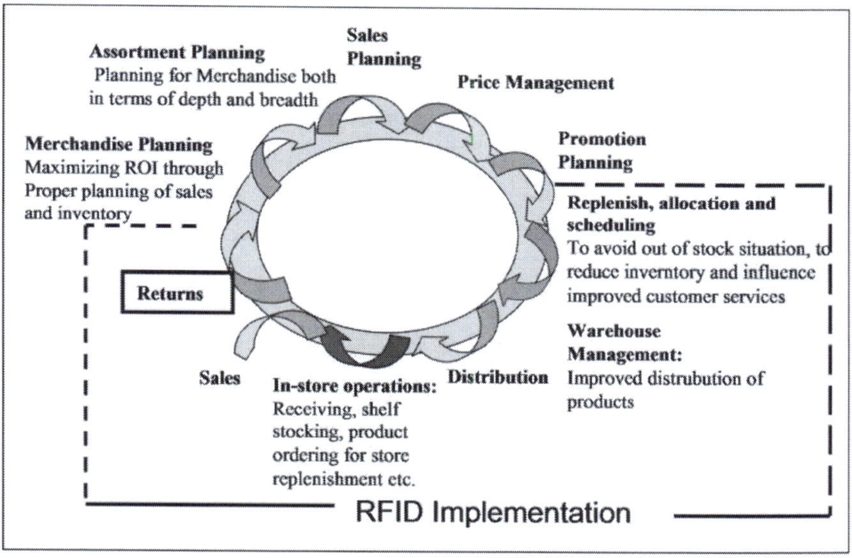

Figure 3. Integrated Supply Chains (adapted from Callana, 2006; Bhattacharya, 2007 p. 14-15)

We see that most of the dominant RFID benefits are focusing on the lower side of the supply chain. RFID technology has the ability to provide up-to-minute information on sales of items, thus can give accurate information about inventory levels. With this accuracy managers may hold their inventory levels at minimum and this may cause to reduce their inventory costs. RFID technology at the pallet level has the potential to automate the distribution of goods between manufacturing plants warehouses and retail stores of different organizations. Companies can cut their costs also down from lost/misplaced inventory (Sounderpandian et al., 2007, p. 105).

There are evidences which prove a positive ROI for warehouse application but in our study we also would like to consider the advantages of in-store applications for store management and customers.

4.1. In-store applications for RFID

In a retail store RFID tag information is generated based on events: A product is leaving a shelf or a product being checked out by a customer at a checkout corner (Sounderpandian et al., 2007, p. 105). The tag readers should be deployed in a shelf; these tag readers are responsible for reading RFID tags of items on the shelf. Items read by the tag at the checkout generate messages for the host system. After processing these messages the host system informs other partner in supply chain. In addition the host system may send some of the RFID transaction data to the enterprise system of the retailer. The host system is connected to the enterprise information system via a virtual private network (Sounderpandian et al., 2007, p. 106).

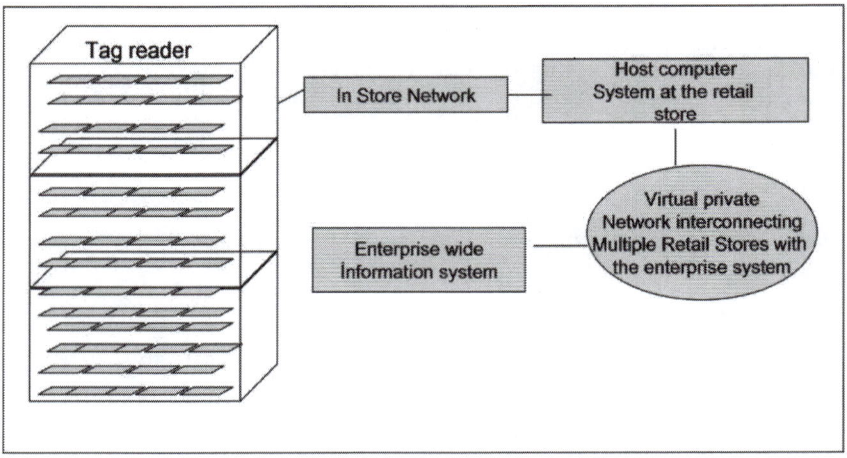

Figure 4. RFID infrastructure in a retail store

Sounderpandian Jayavel; Boppana Rajendra V ;Chalasani Suresh ; Madni Asad M. (2007). Models for Cost-Benefit Analysis of RFID

Implementations in Retail Stores. *IEEE Systems Journal*, Vol..1, No,2 December. p. 106.

In Europe, Metro Group has been using In-store RFID implications in its Future Stores since 2004. At the end of 2008 Metro group has brought 200 sales points in action, included all Metro Cash&Carry big supermarkets,, in 9 central distribution-warehouse of Metrogroup and also most of the Real Warehouses. They also started in 2008 France with RFID applications of 1.3 billion palettes for 89 Metro Cash and Carry Stores.

There are new usage areas for RFID- Instore applications that have been applied by Metro Ag. at Galeria Kaufhof Essen. Applications are following (Metro Group-Future Store- Guided Tour):

4.1.1. Personal digital assistants/smart shelf

Retailers have limited shelf space available. The choice of which items to stock and the allocation of scarce shelf space among the stocked items are relevant issues for the retailer. For individual SKUs these decisions are important determinants of sales and marketing effectiveness. At the aggregate level, shelf allocation is an important factor in the revenue, cost, and eventual profit of a product category. Complementary to the amount of space to allocate to an item, there is the problem of the location of the item on the shelf. For example, items on the lower shelf usually get less consumer attention than items on upper shelves. The items on the lower shelves may therefore have lower sales and may also benefit less from promotions.

Finding the profit-maximizing shelf arrangement while, at the same time, meeting manufacturers requirements is far from easy. A prerequisite to actual shelf optimization is a proper measurement of the effect of shelf layout on sales and marketing effectiveness (Nierop amd Franses, 2008, p. 1).

Retailers are testing the Smart Shelf, where an RFID reader is incorporated into the shelf and stocked with tagged product. The Smart Shelf monitors its rate of depletion, provides an alert when stock runs low and automates reordering to minimize out-of stocks. Gillette is testing smart shelves in an attempt to minimize theft. Because store personnel program the system with store sales data, the system detects behavior outside the norm and can alert store personnel by transmitting information to a personal digital assistant. By identifying the nature of the stock loss and mapping and addressing points of vulnerability, losses in some stores have been reduced by 70% to 80% (Thompson, 2004; p.3).

Advantage to the company with RFID systems: RFID also provides transparency on the sales floor. Every item of clothing is assigned a certain position on hangers or shelves. This data is saved in the outlets database. Employees record these items on shelves with "Personal Digital Assistants (portable RFID readers)". Incorrectly stacked goods could easily be found and resorted (Metro Group-Future Store- Guided Tour; Yalçınkaya; 2007). Advantage to the Customers: In store located RFID readers constantly detect item transponders. Current stock is shown detailed (according to color and size) on the screen. This way customer can check if desired items are in stock on the hangers or shelves (Metro Group-Future Store- Guided Tour).

4.1.2. Check out

Today, staff at the check out scan the barcodes on items to calculate the total amount purchased and deactivate the EAS (electronic article surveillance).

Advantage to the company with RFID systems: During the payment process, data is removed from the RFID system that operates parallel to the merchandise management system. Advantage to the Customers: No links are made between the purchased items and personal data – regardless of whether payment is with an EC Card or credit card. Customer can ask staff to remove the transponders completely if they wish (Metro Group-Future Store- Guided Tour).

4.1.3. Smart mirror

The advantage to the Customers: Customers could check the Smart Mirror to see if the selected garments fit. Integrated RFID readers detect the transponders that are fitted to the clothes being tried on. Detailed information (washing instructions, price etc.) on a chosen product then appears on the mirror's surface if requested (Metro Group-Future Store-Guided Tour).

Also another brand for luxury goods, Prada is using the smart mirrors in its stores in New York Epicenter Store. The mirrors in dressing rooms become magic mirror with a combination of a touch activated display and cameras, so customers may see what they try on from various angles. All articles in this shop have been tagged with RFID transponders which enables this technology (Spektrum RFID, 2011).

4.1.4. Smart dressing rooms

Advantages to the Customers: There were also touch screens in the cubicles in Gardeur shop. Smart Dressing rooms identify which item customers try on and shows product details on the screen. It also gives tips on accessories and possible combinations. Customers can also access details on suggested items by touching screen (Metro Group-Future Store- Guided Tour).

4.1.5. Handheld reader for salespersons

Advantages to the company with RFID systems: Within the store, which has been tagged with RFID transponders a handheld reader tells employees which items are still in stock. This is also an enormous advantage for customer service and availability of goods (Spektrum RFID, 2011).

4.1.6. Anti-theft system

Advantages to the company with RFID systems: RFID readers were also installed at transition points around escalators and lifts. In the future Metro AG is also planning to use passive RFID transponders (Metro Group-Future Store- Guided Tour).

4.2. Cost and benefit analysis of RFID systems in retailing

Benefits of the RFID systems in various industries (defense, healthcare, entertainment ets.) have been widely discussed, but managers still have some concerns about using this technology especially for in-store applications.

The advantages of In-store RFID systems are automatic check-out and reducing inventory costs due to the efficient shelf replenishment. Another issue is also the reduced losses due shoplifting. The points for these concerns are:

- Tag readers cost's, infrastructure costs (hardware and software costs, including the communication network required for RFID implementations)
- Yearly operational or maintenance costs (Sounderpandian, 2007, p. 106).
- Educating sales personnel about the new technology,
- Ethical issues and security concerns from the customers (Bayraktar, Yilmaz, 2010).

Sounderpandian *et al.* (2007) have calculated the formula for a retail store to evaluate whether or not an RFID implication is beneficial for a retail store. Then they showed with the help of a numerical example a retail store which uses part-time employees and implemented RFID system in its stores (Sounderpandian et al, 2007, p. 112): The formula is:

$$FRFID < F + BSL + BPOS + TIC * \sqrt{(1 - VRFID/v)} \quad (1)$$

FRFID = Fix costs: The costs at the maximum number of shelf replenishments $312.000

V= Variable costs: Wages of part-time employees who have been hired depending on the workload and the cost of consumables, $0,75. Variable costs depend on the number of replenishments in a year.

FRFID: RFID implementation costs: Amortized cost of computer hardware, RFID Related Equipment, RFID Tags, salaries of full-time employees, and the wages of (fewer) part-time employees, variable costs are the costs for the consumables.

BSL: Benefit from shoplifting: Using RFID reduces the loss due to shoplifting (for example $10 a day makes $3650 a year)

BPOS= Savings from the use of RFID at POS terminals so the checkout will be faster. Store needs fewer checkout counters and fewer cashiers. BPOS:$ 24.000 (in a year)

TIC: Total Inventory costs: Total costs of carrying and ordering inventory. Carrying costs: The costs of the shelf space: Rent, utilities maintenance of the area where the shelves are kept. TIC: $980.000

Sounderpandian and his colleagues have made the calculation for a retail store and discovered that RFID installation is beneficial.

$$\$750\,000 < \$312\,000 + \$3650 + \$24\,000 + \$980\,000 * \sqrt{(1 - 0{,}10/0{,}75)} =$$

$$\$\,750\,000 < \$961\,805. \quad (2)$$

However in countries where labor costs are relatively cheaper, managers may prefer to pay wages for part-time workers instead of paying for RFID implementation.

5. CONCERNS FOR GLOBAL BRANDS

Another problem for Global Brands is that the franchisees may not agree to tag their products with RFID transponders unless the main company (franchisor) agrees to place the RFID tags in the products before arriving at the retail store. (Number 7 in Figure 5).

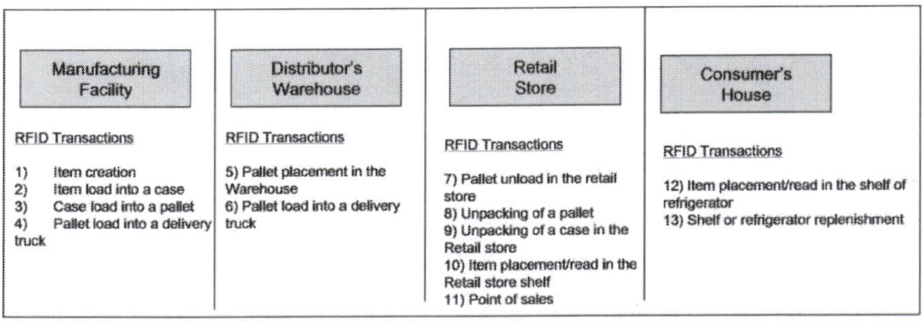

Figure 5. Transition of an item from the manufacturer to the consumer in the supply chain and the relevant RFID transactions.

Sounderpandian Jayavel; Boppana Rajendra V ;Chalasani Suresh ; Madni Asad M. (2007). Models for Cost-Benefit Analysis of RFID Implementations in Retail Stores. *IEEE Systems Journal*, Vol.1, No,2 December, pp. 107.

It is not feasible for a large retailer, one with a daily turnover of 40,000-60,000, to place RFID tags on products before entering the store.

Another issue is that they have to change all labels and price tags three times a year according to the sales etc. Usually on product labels there is information about the country of origin, washing instructions etc. Therefore an extra RFID tag is sometimes too much for an item.

Other problem will be in point 8 and 9 in Figure 5. If the retail store is in a shopping mall, they have to unload their pallets in certain hours, and they can use loading elevators in shopping malls during limited hours. Until the stores open at 10.00 am, all the articles have to be unpacked, tagged, labeled and placed in the store shelves. Therefore with the increase the item numbers have been sold in the store this system could be complicated and costly.

6. CASE STUDY: RFID APPLICATION IN A TURKISH RETAIL COMPANY

Turkey's ready-to-wear clothing industry is one of the major industries of its economy and international trade. According to Sevim and Emek (2006) clothing and textiles have annual sales of $30 billion and a 26 percent share of total export Vol.ume in 2005. Turkey is the fourth largest clothing supplier in the world and second largest supplier to the European Union. Under the World Trade Organization Agreement on Textiles and Clothing this sector continues to maintain and enhance its competitiveness despite the abolition of quotas (Sevim and Emek, Turkish Clothing Industry Report, Export Promotion Center of Turkey, 2006). Germany, the UK and the U.S. are the most important markets for Turkish exports, with export shares of 27%, 18%, and 8% respectively. However, compared to 2004 data, exports to the U.S. have declined 21 percent. Hence, understanding the causal factors has potential for reversing this decline and growing apparel and textile exports to the U.S (Seitz, Neace, Razzouk, Keyfli, Tung,2008; p:173)

Throughout the theoretical findings we prefer to make a exploratory study about using RFID technology. In general, exploratory research is appropriate to any problem about which little is known. Exploratory research then becomes the foundation for a good study (Churchill, 1999;p:103).

Seltiz, C., Wrightsman, L.S.,Cook, S.W. (1976). Research Methods in Social relations, 3rd ed. New York. pp.90-91.

As shown in figure 6, "analysis of selected cases" is a type of exploratory study. The analysis of selected cases is sometimes referred to as the analysis of "insight-stimulating examples." By either label, the approach involves the intensive study of selected cases of the phenomenon under investigation. Examination of existing records, observation of the occurrence of the phenomenon, unstructured interviewing, or some other approach may be used. The focus may be on entities (individual people or institutions) or groups of entities as sales representatives or distributors in various regions (Churchill, 1999; p:113).

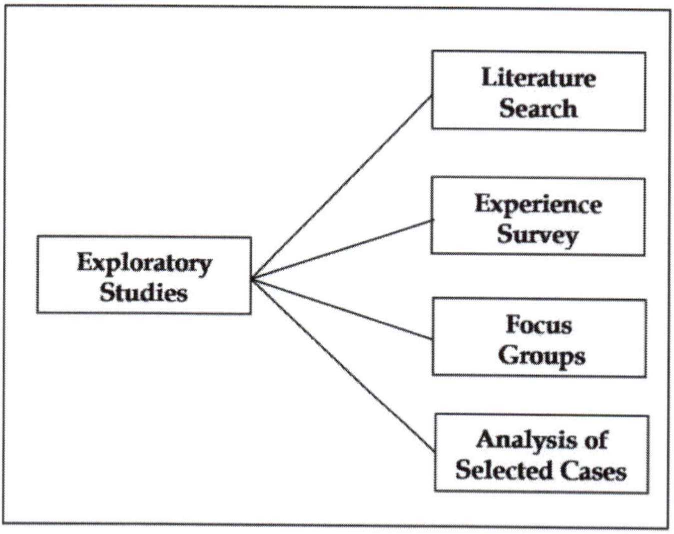

Figure 6. Types of Exploratory Studies

For the case study we have contacted a Turkish retailer and focused on a company. The company (mentioned as X to protect the confidentiality of the company) is one of the fastest growing retailers in Turkey. The company is active in the textile business and its product categories consist of casual children's, men's and women's articles. Their price ranges are between low and medium. After discussing with operations and logistics manager, we gathered enough information regarding RFID.

The company has been using the RFID technology for 3 three years. Contrary to other companies we have seen in the literature, the company X made some tests for its warehouse about the feasibility of the process three years ago but then decided to use RFID technology in its five selected (two stores in Istanbul, three stores outside of Istanbul) stores but not in its warehouse and supply chain system.

6.1. RFID test project

Products should be placed with RFID tags by the suppliers first. For the textile products they have ordered samples for washable, sewn-in RFID tags and made reliability tests for stock control. Companies who tried the string RFID tags had problems because they can break off and get lost easier; therefore, they have built antennas in the warehouses and made RFID track tests for the sample products with sewn RFID tags.

Results of the test: For stock follow ups and inventory levels they have nearly 100% reliability. Defection of the tags and the possibility of not reading the RFID codes within the cases was nearly 0%. In product incoming process into the warehouse:

Defection rate because of the inability of tag reading was again nearly 0%.

For the calculating stock level's accuracy rates were nearly 100% because of the durability of RFID tags. Live time value of RFID tags are 5 years but they should not come near the magnetic areas like near the transformer room etc.

Following up customer complaints: Even if a complaint comes from the customer for a specific product the company can match the Product ID with factory code. Especially The sewn-in tags are more durable than the thermal ones during warehouse processing. This system also facilitates the quality control process.

6.2. Advantages of RFID tags within the supply chain for company X

Unique ID Number: The advantage of unique RFID tags compared to Product ID is each product with the same color, the same size and the same model will have the same Model ID. Even 100 products will have the same Model ID. But with RFID each product will have a unique product ID and company X will follow the products from supplier to consumer with this unique ID number.

Sorting system: Before products have been distributed from the main warehouse, RFID can also make it easier during the preparation stage of delivering goods. Even within the stores, if there is an elevator in the store, building RFID antennas at elevator doors can help sorting the system between the stages and make the follow-up process easier for the employees. Company X delivers 3600 cases/hour daily from its main warehouse to its stores.

Quality of RFID Tags: Company X uses thermal etiquettes now and they can be damaged because of dust, tearing off, getting lost etc. but sewn RFID etiquettes can avoid all these disadvantages.

Collecting of Off-Season Articles: If a company use a trolley with an RFID antenna at the bottom, they can collect all the RFID tagged off-season articles in the store and when the trolley leaves the store and goes to the store warehouse, items can be deleted from the system automatically.

Despite all these advantages the company X decided not to implement RFID technology for their supply chain system at this time. Reasons of these decisions are following:

6.3. Disadvantages of RFID tags for company X in their supply chain

Problems for different Product types: Company X needs different types of RFID tags (for belts, for the coats for the shoes, etc.)

Implementation costs: Each year 160 million products are sold. For each sewn-in RFID tags they had to paid (in 2007) 11-14 cents. If they implement this strategy they would like to eliminate the barcode system.

The cost for 160 million item delivery per year and implementation cost for 320 stores in Turkey (including software costs, RFID tags of all Products, handheld computers for employees, antennas, etc.) is 37 billion dollars.

As fixed cost the company calculated for the stores' antennas at the door and at the check out point.

Variable costs are handheld terminals, RFID tags.

ROI time : In retailing ROI time is approximately one year. According to their calculation company X can get this investment in 27 months.

Production in different locations: The Company has 300 producers and 50% of them are outside of Turkey. Their production facilities are in Bangladesh, Egypt, Sri Lanka and in China. After sewing RFID tags into the products, suppliers should also match the RFID tags with the unique products and load information in to the Network system.

Problems with sewn-in Tags: These tags should be sewn separately and carefully without damaging the antenna. For this sewing process they have to pay 1 cent extra and it makes the RFID tag cost 15 cent per unit. Because the retail prices for their products are not so high, company X can not add these extra costs to their prices.

Different system in the Warehouse and in the stores: If one company decides to use RFID in their warehouses and it they should also implement this strategy in their stores to make a hypenate for the information cycle. If the company uses barcodes in the stores and RFID tags in the warehouses, the information system should always transfer data between Model ID and Unique Product ID. Now in their system, stores are transferring RFID data as a barcode data to the warehouse.

Therefore company X did not implement RFID technology in its supply chain and warehouse system. They are still using barcodes for the product groups in factories and warehouses.

6.4. RFID applications for company X within the stores

Company X selected two test stores for RFID tags in two different areas in Istanbul. They have contacted a Turkish IT company, who invented a special alarm tag EAS (electronic article surveillance) with RFID for them.

Process for the incoming products: Before placing the incoming products on the shelves, sales personnel have to match each product with the barcode and the RFID transponders. In order to simplify this process company X placed a horizontal antenna. Since products are shipped from the factories with a barcode inside, employees should match barcodes with RFID transponders. After the matching process is completed, employees place the products on the shelves.

For missing products in the stores: Company X is not using an intelligent shelf system. If a customer searches for a product which is in store network system but not on its shelf, employees are entering its RFID code in the handheld computers and walking within the store until transponders give a signal for the lost item.

Deactivating RFID tags: After customers finish their payment process, employees tear off the EAS tags (within RFID transponders). These tags and their information are deleted at the end of the day, so one EAS (RFID transponders) can be matched with another barcode and used 10.000 times. With this system customers can be sure about the privacy issue, because they wont carry the RFID tags on the products (Orel, 2006).

6.5. Advantages of RFID tags for company X within the stores

Reducing check out time: Employees are also using this system for the item counting. The main benefit of RFID transponders is reducing check-out time and line. Since the company has low prices, customers usually buy three or more items. Employees are putting all purchased products on the antenna, and the RFID transponder's, system can read them within seconds.

The main advantage of RFID is using RFID reader imbedded baskets in the store warehouse. Employees can bring the items in/from the store with these baskets and RFID readers transport information of these items directly to make stock control processes more efficient.

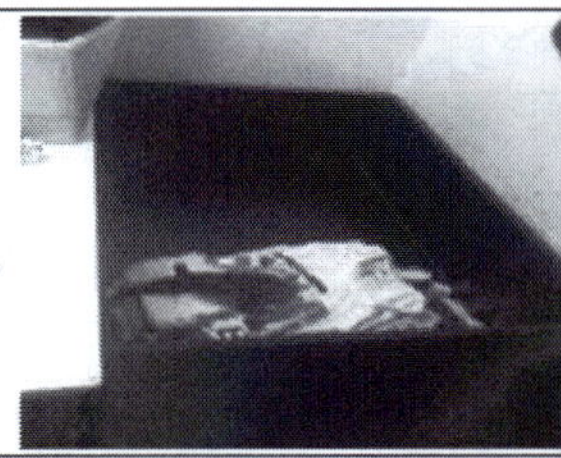

Figure 7. RFID Basket

6.6. Disadvantages of RFID tags for company X within the stores

After discussing with the store manager, she explained the difficulties of RFID tags, as follows;

Using RFID tags in the EAS alarm tags :
Managers ordered these tags in order to decrease number of etiquettes which have been tagged on the items but these tags are really big and cause some difficulties for small items (like baby products, underwear). They damage the item or the item is so thick, alarm can fell down easily.

Figure 8. Difficulties of RFID tags

The impact of RFID for Incoming and Store Warehouse Process:

This test store is one of the most crowded stores of the company. Each day this store receives 120-150 big boxes of products.

Security Issue:

The needles of the alarms are very short and can be taken off very easily. This problem increases also shop-lifting per day (especially for expensive products like coats). To eliminate this problem Company X hired extra security personnel (4 full-time, 2 part-time). One person is responsible for the entrance and the others are responsible for inside the store and for dressing rooms. But still the Company can not prevent this situation.

For the incoming process employees should open the boxes. Since each product is placed in a bag they should open them before placing the incoming products on the shelves. Sales personnel have to match each product with the barcode and the RFID transponders. Than place the tag into the products. This process negatively effects the time for bringing the product on the shelves.
This process takes 4 seconds. If employees would eliminate the matching process (in case of using the barcodes only) they only have to read the barcodes per product and this reduces the process by 1.5 seconds. These 4 seconds with RFID tags could be also longer sometimes if one tag is defect or if RFID tag has been matched with the wrong barcode.
This store receives 1000 products daily and with RFID tags they need 4 hours to finish incoming process and two employees are full time dedicated to do the matching.

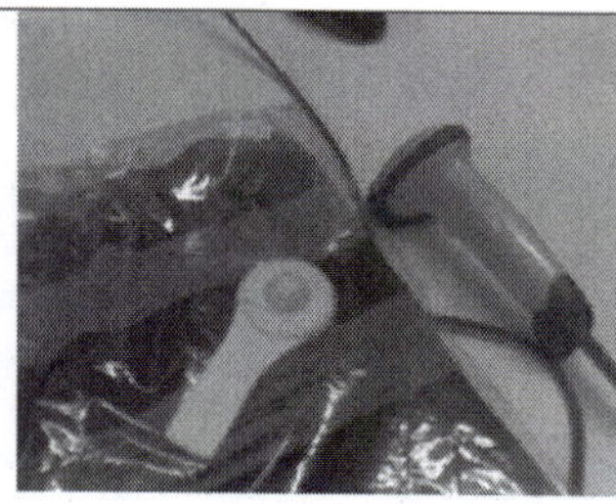

Figure 9. The impact of RFID

Normally reading the barcodes of purchased items separately takes for multiple items 52 seconds in average. Reading these items with RFID readers decreases this time by 8 seconds. But RFID readers are located under the cash register (see picture at the right side) and the radius of these readers are 60 centimeters or more. Therefore sometimes the reader may not read only the tags of the purchased item but extra tags around the reader as well (like the tags which have been taken off and thrown in to the basket like in the picture).
In this situation the employee should repeat the purchasing process from the beginning and this decreases check-out point performance of the store.

Figure 10. Check-Out Point

Technological support service from the RFID Company:

In our case Company X has an outsourced RFID company, which supplies them the tags, portable readers, check-point readers etc. But this company was not able to give appropriate service for RFID. Tags, readers can be damaged easily and RFID company charged each time when they have to fix the devices extra cost to the head office of Company x.

Problems from the customer point of view:

Because of the reading problem at the check-out point customers may sometimes pay for the products that they had not bought. The RFID reader may read a barcode multiple times. The percentage of this problem is 5% to 6%. But if a customer has bought ten items, it can take half an hour for the employee to renew the process. Also for EAS alarms there is a last control device at the check-point before customer leaves the store. But for RFID there is no device and customers may face with alarm at the exit if one tag remained on the items.

These problems damage also the customer relations of the Company X.

7. CONCLUSION

For the new technological systems like RFID, mobile etiquettes and printers etc. retailers want the ROI less than one year. If the ROI time is longer than one year, they prefer not implementing these technologies.

To also consider the opportunity cost effect many retailers would like to invest this amount into advertising or other short term investments and gain the information instead of RFID with other cheaper technologies.

For in-store usage RFID Companies have to work especially on security issues. During the R&D process they have to work closely with the retailers and store managers in order to find the best solution, which will satisfy the need of each member in the supply chain.

If the retailers would like to implement this technology in their stores, it is appropriate only for stores which receive 100 items (on average) daily, have expensive merchandise and relative low in-store traffic.

But still the advantages are really very important. Also a lot of companies are ready to invest in this technology as soon as the cost for RFID tags decreases. Therefore for future research into new uses and of RFID tags need to be investigated.

REFERENCES

1. R. Angelles, 2005RFID Technologies: Supply Chain Applications and Implementation Issues, Information System Management, Winter 2005, 5165

2. B. Bacheldor, 2006Forecast Indicates Strong RFID Demand by Heavy Manufacturing. RFID Journal, Available at: http:// www.rfidjournal.com/article/articleview/2652accessed on 23.02. 2011)

3. C. Bardaki, K. Pramatari, 2008IP-Mapping a RFID-integrated Shelf Replenishment Information System for the Retail Industry to Assess Information Quality. Proceedings of the 16th European Conference on Information Systems (ECIS), Galway, Ireland, June 911

4. A. Bayraktar, E. Yilmaz, O. Yamak, 2010Implementation of RFID Technology for the Differentiation of Loyalty Programs. Journal of Relationship Marketing, 9,1:30-42.

5. M. Bhattacharya, C. H. Chu, T. Mullen, 2007RFID Implementation in Retail Industry: Current Status, Issues and Challenges; Decision Science Institute (DSI) Conference, Phoenix Arizona AZ, Available at: http://citeseerx.ist.psu.edu/viewdoc/download?doi=10.1.1. 77. 2715&rep=rep1&type=pdflast accessed on 26.02.2011).

6. G. Callana, 2006Improving Supply Chain Efficiency in Retail Sector. Supply Chain Europe, London: May/June. 15Iss.2; 22

7. G. A. Churchill, 1999Marketing Research: Methodological Foundations. Dryden Press Seventh Edition. Orlando.

8. E. C. Jones, C. A. Chung, 2008RFID in Logistics: A Practical Introduction, CRC Press Taylor & Francis Group, USA.

9. B. Jabjiniak, G. Gilbert, 2004RFID Warrants a Strategic Approach, Business Integration Journal, 2931

10. J. Maeda, 2006Laws of Simplicity : The MIT Press, Cambridge Massachusetts London England. 100262134721

11. Metro Group-Future Store- Guided Tour: Available at: http://www.future-store.org/fsi-internet/html/de/23786/index. html; http://www.future-store.org/fsi-internet/html/de/1613/index.html, last accessed on March 2011.

12. E. V. Nierop, D. Fok, P. H. Franses, 2008Interaction Between Shelf Layout and Marketing Effectiveness and its Impact on Optimizing Shelf Arrangements. Marketing Science, Articles in Advance, Published online ahead of print June 23, 2008 119

13. A. Sabbaghi, G. Vaidyanathan, 2008Effectiveness and Efficiency of RFID technology in Supply Chain Management: Strategic values and Challenges. Journal of Theoretical and Applied Electronic Commerce Research 0718-1876Electronic Version 32August, 7181

14. V. Seitz, M. B. Neace, N. Razzouk, E. Keyfli, C. W. Tung, 2008Turkey: gaining market share in the U.S. ready-to-wear clothing market. Review of Economic and Business Studies, 1171178

15. C. Seltiz, L. S. Wrightsman, S. W. Cook, 1976Research Methods in Social relations, 3rd ed. New York. 9091

16. Sevim and Emek, Turkish Clothing Industry Report2006Export Promotion Center of Turkey,

17. Spektrum RFID: Metro Group: Future Store Initiative, available at (http://www.future-store.org/fsiinternet/get/documents/FSI/ multimedia/pdfs/broschueren/WISSB_Publikationen_Broschueren_S pektrumRFID.pdf)last accessed on February 2011.

18. Jayavel. Sounderpandian, Rajendra. V. Boppana, Suresh. Chalasani, M. Asad, Madni, 2007Models for Cost-Benefit Analysis of RFID

Implementations in Retail Stores. IEEE Systems Journal, 12December, 105114

19. O. Thompson, 2004Supply chain payoffs with RFID:With advances in technology and increased Volumes, RFID will become cost effective for many applications. Food Engineering Magazine. April 6 2004 (http://www.foodengineeringmag.com/Articles/Column/718f8b49082f8010VgnVCM100000f932a8c0

20. F. D. Orel, 2006Mağazacılıkta Ümit Veren Yeni Bir Teknoloji: RFID (A new technology which gives hope to retailing), available at http://www.fatmaorel.net/bizim%20market/RFID.pdf,last accessed on 14 March 2011.

21. L. Yalçınkaya, 2007RFID ve Hazır Giyim Sektörüne Katkıları (RFID and its contributions to ready to wear industry), March, available at http://sts-rfid.com/docs/doc1.pdf,last accessed on March 2011.

CHAPTER 8

A Computational Architecture Based on RFID Sensors for Traceability in Smart Cities

Higinio Mora-Mora *, Virgilio Gilart-Iglesias, David Gil and Alejandro Sirvent-Llamas

Specialized Processors Architecture Laboratory, Department of Computer Science Technology and Computation, University of Alicante, 03690 Alicante, Spain

ABSTRACT

Information Technology and Communications (ICT) is presented as the main element in order to achieve more efficient and sustainable city resource management, while making sure that the needs of the citizens to improve their quality of life are satisfied. A key element will be the creation of new systems that allow the acquisition of context information, automatically and transparently, in order to provide it to decision support systems. In this paper, we present a novel distributed system for obtaining, representing and providing the flow and movement of people in densely populated geographical areas. In order to accomplish these tasks, we propose the design of a smart sensor network based on RFID communication technologies, reliability patterns and integration techniques. Contrary to other proposals, this system represents a comprehensive solution that permits the acquisition of user information in a transparent and reliable way in a non-controlled and heterogeneous environment. This knowledge will be useful in moving towards the design of smart cities in which decision support on transport strategies, business evaluation or initiatives in the tourism sector will be supported by real relevant information. As a final result, a case study will be presented which will allow the validation of the proposal.

KEYWORDS

RFID smart sensor network; cyber physical systems; communication technology; computational architectures; smart city; track and trace system

1. INTRODUCTION

Technological progress in recent years has made it possible to extend the use of Information Technology and Communications (ICT) to new applications with the objective of improving citizens' quality of life. This idea has been of increasing significance in the political agenda as well as in the public services area. In the cases where there is no planning of development consistent with the efficient and sustainable management of urban resources this can start to cause real issues.

It is recognized that to either obtain a common definition of Smart City (SC) or to define global trends for SC is complicated. There are several reasons that contribute to this. On one hand, cities are complex systems typified by massive numbers of interconnected citizens, a lot of businesses, various modes of transport, communication networks, including services as well as utilities. On the other hand, this leads to a big diversity of technological, social, economic and organisational problems which expose the environmental sustainability of cities. Besides, there is no total consensus about the significance of SC but there is wide agreement concerning the fact that SCs are characterised by the use of ICTs. The goal is, in general, to make better use of resources in cities [1]. There are numerous examples which we could cite of cities that are supplementarily prepared with ICT systems in order to meet this objective in the fields of energy, telecommunications, ecology or transport [2,3,4,5]. Improvements in the management efficiency of these elements are directly perceived by citizens as quality of life improvements.

These reasons have led us to work on the citizen track and trace problem in urban environments. Progress on this issue may provide relevant information for decision-making in many of the above problems related to the management of a smart city, such as in the planning of transport infrastructure (roads, subway, bus lines) or modification of existing ones, in the construction of new public buildings (schools, hospitals, museums, sports areas), in designing tourist routes through the city, etc. Other uses of the obtained information can contemplate economic objectives such as a commercial rating of avenues and streets or the design of tourist offerings in the city.

Moreover, building increased intelligence into things around us and connecting them to the network to transfer data without requiring human interaction is a current trend of technological development called the Internet of Things (IoT) that allows us to meet and interact better with the surrounding environment [6,7]. IoT technologies such as sensing, computing and mobile communication could be employed to collect data on the functioning of the city. In this respect, how the data of interest is collected is the major challenge in most cases. Once obtained, another important issue is how to connect sensors to the web [8] to enable provision of cloud-based management and services. In this field, the development of distributed sensor infrastructure is one of the prominent research areas related to IoT [9,10].

In the work described in this paper, a method and a computer architecture to determine the movements of people in the city is proposed. The research done takes into account the current communication technologies, the ease of implementation and deployment and key aspects about user privacy. The main scientific contribution of this research is the integration of several new technologies and solutions of IoT to build a comprehensive system to track and trace citizens in city working environments. As a result, smart sensing technologies with processing and communication capabilities are proposed to produce cloud services for citizen traceability.

The remainder of the paper is organized as follows: Section 2 covers problem analysis, related work and the research methodology used; Section 3 presents the methods to know citizens' movements with its dominant distributed components and decisions taken; Section 4 discusses aspects of communication and structuring of location information and describes the centralized architecture; Section 5 describes the main characteristics of the services provided for using the generated information; next, Section 6 describes the implementation based on a case study in a university environment to check the correct performance; finally Section 7 concludes with a discussion of critical outstanding issues and future research directions in this area.

2. POPULATION TRACK AND TRACE ISSUES

2.1. Problem Definition

Track and trace is a term frequently used by delivery and logistics companies [11,12,13,14,15]. In the context of items and products, track and trace implies a process of determining their traceability understood as the current and past locations (or even more information) of a single item. In this work, the main idea is to use this general approach in a totally different context regarding people in order to discover and determine citizen movement flows. According to this definition, the general problem addressed in this paper focuses on achieving structured information about the movement habits of citizens in the geographical area of a city.

The main challenge to be resolved is the way in which the population can be traced preserving their anonymity, without necessarily requiring their collaboration, and in non-controlled and heterogeneous environments. In addition, the processing and centralization of information obtained must be considered to be useful for decision-making, and it should be aware that the number of citizens can be very large, even considering a small urban geographical area of an average city.

2.2. Literature Survey

The citizen track and trace issue is not new, as evidenced by the amount of research on this topic. Indeed, there is some concern about maintaining user privacy due to the proliferation of systems and high integration rate of mobile devices connected in the hands of citizens [16,17]. However, the benefit of knowing citizen's traceability information for the design of cities has spurred studies and works which try to determine it while preserving the privacy of citizens.

Transport planning is an example of a task where an analysis of human and traffic flows can be very useful. This issue is essential since the populations of cities are increasing dramatically. Decisions regarding this have been studied for many years. As a result, there are numerous proposals for successful decision-making processes in transport planning [18,19,20,21]. Such decisions involve many actors such as members of the city council, representatives of the transport companies and even the citizens. Usually, council members along with the company representatives come together to form a regional consortium. Therefore, this consortium is the scenario where the decisions regarding public transportation are made. Setting the routes of the various services such as buses or subways is one the main decisions they have to perform. Due to this fact, they must analyse data about the city and its citizens in order to extract the knowledge about their habits and transportation necessities, i.e., the analysis of human movement trends in urban areas. However, although these guidelines provide a well-studied framework for decision-making on transport, there are some critical shortcomings that have been detected in some of the current literature [22,23].

First of all, very often there is an issue of the obsolescence of design plans. Traditional planning processes can become obsolete when cities grow significantly. If too much time passes between when a project is planned and it is executed, by then it will be outdated and certainly will have lost some of its usefulness. To address this problem, there are works proposing a system based on the use of SC tools that allow projects to evolve dynamically with the possible changes in the scenarios over time [24].

Secondly, there is another important issue concerning the procedures used to get the data and especially how can they be used for providing decision support. Traditionally, the information used in these processes is obtained through field studies (e.g., manual recounts of people and vehicles, census, etc.) along with information obtained from interviews with the different actors involved, namely, citizens, transportation companies, and event managers, among others [22]. However, these massive surveys and interviews are very expensive, thus they cannot be performed frequently. Due to this fact, they rapidly become obsolete and if we trust such outdated information we will not take into account the real situation. Furthermore, these data sources lack information about individual trajectories and consequently we cannot compute the real individual movements of people [22,23], but rather we are able to obtain a global and a simple view of citizens' movements.

It is highly desirable, therefore, to use technology to effectively determine the human traffic flows. The main technology created for this purpose is the Global Positioning System (GPS) [25]. With GPS, we can precisely and easily establish our position on the world map, so it is used for orientation purposes by drivers and pedestrians since it has become quite affordable. In addition, it is now integrated into many devices such as mobile phones or other ubiquitous devices [26].

An example of an application for this problem can be found in a study conducted in the cities of Milan (Italy) [27] and Beijing (China) [28]. In the case of the Italian city, the data come from GPS devices installed in a rental car fleet. In the Chinese city, the GPS devices were installed in a taxi fleet. These papers present some analytical methods as well as data mining tools in order to provide discovery knowledge processes. They rely on a central node responsible for processing all the information acquired from the GPS devices.

Based on the information obtained about the trajectories, there are several studies which focus on data analysis and manipulation to extract information of interest. In this regard, query languages and tools for trajectory analysis oriented to query trajectories computed from GPS devices have been established [29]. These researches develop a clustering technique that allows extracting traffic patterns over indicated periods of time. The results of these analyses allow one to trace road communication routes much faster than existing patterns and the results obtained by tools like Google or Bing maps.

However, the use of GPS raises some issues to consider. One of the most important of them is the transparency of use: access to the positioning data of users requires that they allow it, either by configuring or installing an application designed to carry out such data collection on their mobile devices. This configuration involves extra effort that not everyone may be willing or able to undertake. Moreover, if it is activated, the user will be aware that is transmitting data at all times. Therefore, it becomes clear that it is not a transparent mechanism for the user. In addition, there is an added problem: the preservation of privacy [30]. By knowing the exact location of people in real-time and the trajectories they have followed over time, it is possible to study their movement habits and even predict their location at a certain time of day. In the wrong hands, such information may compromise not only the privacy of citizens, but also their safety. For instance, it would be possible to know with precision at what a time a citizen is not at home in order to commit a theft. Finally, besides the two mentioned problems, an additional problem is the power consumption since devices require their own power source to operate in addition to the cost of the devices themselves. Despite being integrated into many medium and high range mobile phones, they are still relatively expensive if purchased separately.

The use of the installed wireless infrastructure, has also been considered for this application [31]. This includes mobile phone networks and wireless local (Wi-Fi) and wide (WiMax) area networks. These networks were not designed for this purpose but to rather access the Internet and communications services offered by telcos. However, from the moment in which each base station is aware of the presence and identity of each connected device, it is possible to offer tracking operations while the user is moving from one station to another [32,33]. Some research also includes filtering and interpolation operations to predict where the user is located [34,35,36]. However, due to their scope and current coverage, they cannot provide accurate situation information about where the user has passed. Moreover, placing a greater number of stations for this purpose could significantly increase the infrastructure cost.

Mobile telecommunications technologies (GSM, 3G, LTE, etc.), can also be used to perform large-scale tracking and tracing of users, however, they are not considered in this work due to the lack of necessary precision and the need of having a mobile terminal. Although their degree of penetration in the population is very high, their use for this purpose involves other drawbacks about cost and user privacy. Other alternatives such as CCTV, are discarded due to the technical difficulty of developing automatic methods for tracking individual users in uncontrolled contexts and the fact it can cause a stronger invasion of privacy due to the feeling of being "watched" by surveillance cameras.

An alternative to previous techniques for sensorization of citizen movements is to use Radio Frequency Identification (RFID) communication technology [37,38,39]. This technology has also been widely used to obtain location information of users and objects. Table 1 summarizes the main recent works on this issue as a representative sample. All the proposals analyzed work in controlled environments, that is, the infrastructure deployment has been designed previously and provided by the system managers. When the users access a system are given a tag to carry around the working environment.

Many studies have been conducted for track and trace indoors and outdoors. For instance, outdoors, the City of London introduced a system for data collection in its transportation system including buses, subways and trains [40]. The idea was to track users of the transport network through payment and access cards equipped with this communication technology. With the information gathered it was determined the level of use, number of travelers and public transport habits in the city and to know the stations, breakpoints, origin and destination of network travel streams. For indoors, systems to know citizen movement flows have also been implemented [41], for instance, visits per museum [42]. With the results, visit patterns of the different exhibits and areas were obtained, participants with atypical behaviours were identified and even relationships between visitors groups were established.

Table 1. RFID-based track and trace works.

Work	Main Application	Working Environment		Tag Deployment	Reader Deployment
		Indoors	Outdoors		
RFID enabled supply chains [11–14]	Traceability services	X	X	Passive tags embedded into objects	Fixed or mobile readers. Along supply chain
RFID delivery system [15]	Intelligent Transport Systems		X	Passive tags embedded into objects	Fixed or mobile readers. Along delivery system.
RTSV [43]	Tracking System for Vehicles		X	Passive tags installed on cars	Distributed throughout the city.
The London Oyster Card Data [40]	Public transport planning	X	X	Passive tags Inside transport cards	Installed on entries and exits of transport system.
iWalker [44]	Assistance services of location and obstacle detection	X		Placed anywhere in the environment	Embedded into walkers
Tracking science museum [42]	People Traceability	X		Passive tags on nameplates carried by users	Distributed throughout the floor
SIP-RLTS [45]	Location Tracking System	X		Passive tags carried by users (patients)	Readers carried by workers (medical)
LANDMARC [41,46]	Indoor location sensing	X		Active tags on grid array deployment	Distributed throughout the environment
Blind User [47]	Location and Proximity Sensing	X	X	Passive tags; Indoor: Grid array deployment over floor; Outdoor: Along edge of the sidewalk.	Readers carried by users
Pervasive mining [48]	Tracking people in a pervasive mining environment.	X		Passive tags carried by users	Distributed throughout the environment
Privacy-Preserving Solution [49]	Tracking People in Critical Environments	X	X	Passive tags carried by users	Distributed throughout the environment
Social interaction [50]	Person tracking	X		Passive tags carried by users	Distributed throughout the environment
Cameras and RFID [51,52]	Tracking and identification people	X		Passive tags carried by users	Distributed throughout the environment
RFID Inside [53]	Tracking and identification people	X	X	Passive tags inserted in users	Readers carried by workers (medical, security, *etc.*)
Tagging Demented Patients [54]	Tracking and identification people	X		Passive tags carried by users	Readers carried by workers (medical)
WSN and RFID [55]	Person tracking	X		Passive tags installed on objects or people	Distributed throughout the environment
Elderly Living Alone [56]	Person tracking	X		Passive tags carried by users	Distributed throughout the environment
Peer-to-Peer Networks [57]	Location Tracking System	X		Active tags carried by users	Distributed throughout the environment
REACT [58]	Children location		X	Passive tags carried by children	Distributed throughout the environment

RFID technology overcomes some of the disadvantages described for GPS, as it does not require too much cooperation from the user and reduces energy costs. However, its use for tracking does not provide exact locations but, like other wireless networks, the temporal location marks are limited by the locations of the signal reader antennas and the analysis of the Received Signal Strength Indicator (RSSI) of the readings [56]. Thereby the resolution of the trajectories followed by the users will be lower than in the case of the GPS. The outdoor scope of this technology is also not comparable to GPS coverage. There is a shorter range variation named Near-Field Communication (NFC) [59] that requires much less power to operate. Moreover, passive RFID/NFC technologies may be present in a multitude of objects and devices that people wear [38]. In many cases, they are integrated into standard cards made of plastic with identification and access functions [60], credit or debit cards [61,62,63] or even paper tags on clothing [64,65,66,67], however the functions for which these elements were designed were not tracking people in itself, but rather other applications such as localized identification, payment processes or object tracking in stores.

After the review carried out, it is observed that the activities of localization and tracking can provide valuable information to improve infrastructure development and user satisfaction for a variety of applications. From all the reviewed literature on track and trace systems, the most cited solutions show significant drawbacks. The Global Positioning System is the most suitable and it provides better results due to its precision and accuracy, however, this implies the necessary participation of citizens for track and trace activities. The citizen must have a device with computing and communication capability that enables it to provide the captured information, usually a mobile phone on which an application has been previously installed and configured. These requirements consume a considerable amount of energy that may affect the autonomy and the availability of the device. In addition, these solutions involve knowledge of personal data of the citizen like the terminal number. These features imply that the system is restricted to a limited set of the citizens, so the sample would be not representative to provide good service for managing the resources of all citizens. Elderly citizens who are not familiar with ICT, citizens who do not have access to such technology or citizens unwilling to provide personal data, among others, would not be represented in the datasets. Moreover, these proposals are only valid for open city environments, which would imply the inability to track citizens in indoor environments where GPS technology is not adequate. Nevertheless, these spaces also require efficient management of their resources.

To take advantage of the extensive set of connected mobile devices as well as the installed infrastructure, the use of wireless networks is another option in which proposals have already been made. However, the results it gets are not very precise and it also produces privacy intrusions as the identification of the user is known.

Another alternative that can be provided is the use of RFID/NFC, a technology increasingly found embedded in devices and other portable fixtures accepted by the public. For its functioning, not only energy consumption, but also expensive equipment, are not required. In this case the main obstacle is that the technology is underdeveloped and in general, antenna networks prepared for this purpose are not available.

Most related works that use RFID technologies are focused on tracking systems and only few proposals have contemplated both the track and trace problems. Most of them apply their solutions in closed and limited environments, where the characteristics of reliability, service quality and scalability are not necessary. However, the development and proliferation of this RFID technology among citizens during the last few years, opens up new possibilities for its use in track and/or trace tasks based on user-centric schemes. The characteristics of this technology may allow mixed deployment both indoors and outdoors. In the analysed work, the use of RFID technology is limited to reading the homogeneous tags provided by the system owner and sending information to a central system, without prior processing, storage or filtering, as these tasks are delegated to a central node, which can generate a bottleneck.

It is noteworthy that none of the literature proposals focus on the provision of the information collected to third parties, in a dynamic, personalized and open way. For this purpose, the development of cloud-based architectures may represent a good choice for the design of flexible management architectures and service providers. This aspect could enhance the development of decision support systems for the efficient management of resources and oriented to meet the needs of citizens. The scope of the limitations and advantages mentioned will be discussed in the following sections to lead to the design of the proposed architecture according to the objectives.

3. RESEARCH METHODOLOGY

The working hypothesis of this research is that it is possible to know and process citizens' movement flows across the city by using new IoT technologies.

Once the problem has been described, the hypothesis established and the related work analysed, it is necessary that the comprehensive track and trace system becomes clearer. This system should allow the acquisition of locations of citizens in heterogeneous and uncontrolled environments, indoors and outdoors, to facilitate the transparent incorporation of citizens into the system. Furthermore, the information acquired should enable providing information about citizen movement flows to someone else for efficient management of the resources of cities and to be able to provide better services to these citizens.

In order to develop our proposal, a research methodology based on business process management [68,69] using the Eriksson-Penker formal notation [70] it has been followed. Process management is a strategy for structuring a complex process into a sequence of tasks that can be shown as actions that transform inputs into some other output elements. This transformation must be aligned with previously defined goals, which are considered as strategic, to meet some needs or gaps identified in the problem context. In this way, the defined process will achieve the object system of this work in a systematic way, selecting the most appropriate techniques and tools to meet the objectives in order to solve the identified problems. In the proposed methodology, each of the tasks identified represent a stage of the research, and therefore these will have associated one or more scientific methods as described below.

Specifically, from the starting hypothesis, Figure 1 shows the main process carried out in the research, named Design of citizen track and trace system, where the input element («input») represents current track and trace systems. This «input» must be transformed through this process into a novel citizen traceability service («output»), addressing the needs and gaps identified previously, which now are represented by the strategic objectives to be able to guide the research process («goals»). To perform the transformation to achieve the objectives established, the controllers («Control») and facilitators («supplies») which are required to guide this transformation must be identified. Particularly, these elements represent strategies, paradigms, techniques and technologies to be integrated into our proposal. In order to make easier its identification and, therefore, make clearer the problem, the main process has been divided into three sub processes or tasks.

The resulting sub-processes are the following (Figure 1): first of all, the sub-process Design of the method for citizens' location acquisition, focused on establishing necessary technology infrastructures to get citizen data. Secondly, the sub-process Design of the method of communication and structuring of citizens' location, involved in ensuring the correct reception of the citizen location data. Finally, the sub-process Design of the method for citizens' flow provisioning focused on provisioning standard citizen flows to third parties.

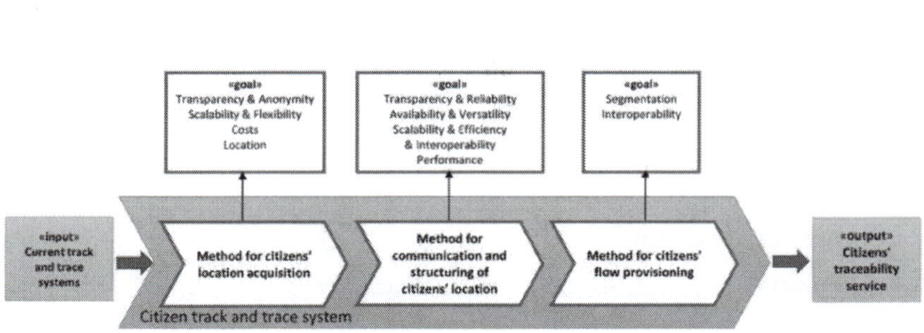

Figure 1. Modelling with Eriksson-Penker notation of the process of obtaining citizen movements flows.

The result of this process will be a computational architecture in order to read and process citizens' movement following the overall scheme shown in Figure 2. The final design of this architecture will be led by the decisions made in each sub process of Figure 1. The research made, uses the knowledge and experience of the research group in the field of designing specialized computational architectures [71,72,73].

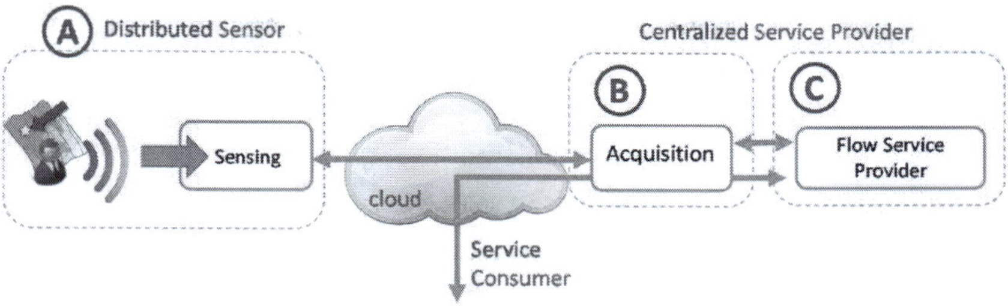

Figure 2. Overall Computational architecture of the citizen track and trace system.

As can be seen, it is composed of a distributed module and a centralized one. The distributed module (A) implements the reading of user location data. The centralized module (B) acquires the data from sensors and provide to the consumers (C) the citizen traceability service to help when making decisions on urban management. The functioning and components of those modules will be detailed progressively in the next sections.

3.1. Method for Citizens' Location Acquisition

This task is focused on the identification of the most suitable location technology and the location infrastructure design as a starting point for addressing the problem to achieve the following goals, as noted in Figure 1: gathering of geographic information (location) should be transparent to the citizen, meaning without requiring their cooperation and without their intervention in managing and setting the devices, without affecting its economy and independently of whether the citizens' transits through both indoors and outdoors environments (transparency); the designed method will have to maintain the user anonymity and protect personal privacy, avoiding any capture of citizens' personal and confidential information (anonymity); the technology used should allow its functioning and quality of service (QoS), regardless of the number of citizens interacting with it and the size of the area that will be monitorized (scalability) and the associated infrastructure should provide agile adaptation to any environment (flexibility); finally, the deployment of the necessary infrastructure should have a low financial and consumption costs (costs). With these design goals in mind the intention is to facilitate the placement of a valid infrastructure and minimize citizens' rejection. Figure 3 shows a complete schematic of this process following the methodology criteria.

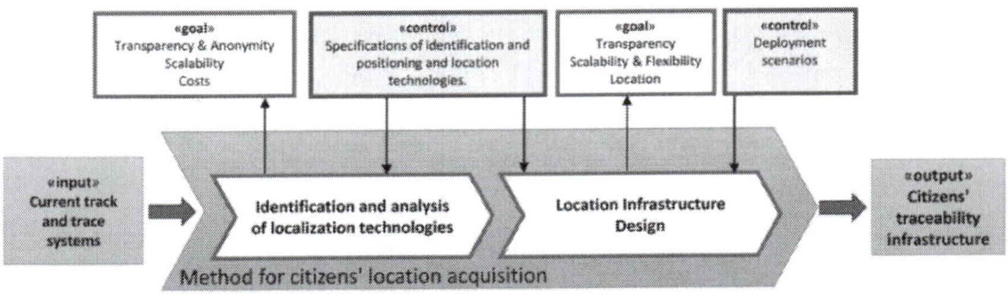

Figure 3. Modelling with Eriksson-Penker notation of Design of the method for citizens' localization acquisition process.

3.1.1. Identification and Analysis of Localization Technologies

For the selection of localization technology, we take into account the analysis and the experience gained in the review of other work and the performance and benefits that are currently available [25, 26, 30,34,35,36,38,42,59,74,75,76,77].

Table 2 summarizes the main technical characteristics as they relate to the proposed objectives. Regarding the objective of transparency, obtaining positioning with GPS systems or Wi-Fi networks do not offer transparency and the desired anonymity for such applications, since in both cases the collaboration as well as users' permission is required. Specifically, it is necessary that a citizen has a device that supports these technologies, he must install and configure an application designed for this purpose, provide the permission to transmit location information, which may include personal information, and, for all this, he must have a minimum knowledge and technological skill. In addition, the accuracy and coverage of GPS could be a privacy and legality concern, in terms of providing more than the location information necessary for the purpose of the proposal, which is the efficient management of city resources. Besides, although other wireless technologies (Wi-Fi, GSM, etc.) could determine the location, the accuracy to address the problem described would not be valid in all cases. All this could hinder the acceptance of the proposal by the general population and therefore it would not be possible to obtain representative samples for the established purpose.

Table 2. Main features of communication technologies.

Technology	GPS	RFID/NFC	Wireless Networks
Main architecture	Triangulation positioning by satellite.	Set of antennas or readers and receivers.	Set of antennas or readers and receivers.
Communication	The user receiver obtains the signal from satellites and calculates the position.	Readers inspect receivers to determine whether they are present.	Receivers report that they are present.
Operating frequency	1100 MHz to 1600 MHz	Active: 455 MHz, 2.45 GHz, 5.8 GHz; Passive: 128 KHz, 13.6 MHz, 915 MHz, 2.45 GHz	Wi-Fi: 2.4 GHz, 5 GHz; WiMax: 2.3 GHz, 3.5 GHz; Cellular mobile: 800 MHz, 1900 MHz, others.
Cover	Worldwide; Outdoor environment	Depending on antenna network deployed. Outdoor and indoor environments	Depending on antenna network deployed. Outdoor and indoor environments.
Range	Worldwide	Active: ~100 m; Passive: 0 to few meters	Wi-Fi: 30 to 100 m WiMax: ~50 km; Cellular mobile: ~35 km
Power consumption	Very high	Passive tags receiver: Very low.	Very high
Deployment Costs	Satellites: Already deployed and free to use; Present in mobile devices	Need to deploy readers network. Present in mobile devices and other everyday user accessories.	Need to deploy antenna network. Present in mobile devices and other user accessories.
Localization	Receiver position.	Antenna position.	Antenna position.
Transparency and Anonymity	Low	High	Low
Usual application	Navigation, topography, land levelling, *etc.*	Identification, access control, payment, *etc.*	Internet access and communication services.

It is true that in terms of cost or ease of deployment of the infrastructure, there is already a high degree of integration of devices with GPS or Wi-Fi connectivity among the population. However, the energy cost of their use is high and its use may only for track and trace reasons not be justified. Finally, all the technologies are scalable to the deployment of infrastructure used, although the GPS positioning has an advantage since it does not require additional elements apart from the satellites. However, in the domain of the problem studied, the use of GPS is limited in indoor areas which should also be analyzed to manage their resources.

For all these reasons, among the analysed alternatives our proposal is to use RFID communication technology to track and trace citizens. As mentioned in the previous section, these technologies are having an expanding and continuous implementation and currently many devices incorporate it [78]. This allows the use for this purpose of the tag park that is becoming larger and it is currently already installed in many applications such as payment cards [63], clothing [64,65,66,67], access to transport services [40,62] or tourism [60].

In most previous applications the user's identity is not known and therefore it transmits only the identifier of the RFID tag worn by the user (in most cases, the tag could be attached to interchangeable everyday elements such as an umbrella or book which could be carried by different users). Consequently, the system doesn't know the correspondence between citizens and RFID tags. This technology does not allow a user to refrain from providing the location, since he or she will not be able to tear off the RFID tags from their clothes, accessories or cards. In that sense, using this technology for traceability should be treated similarly to CCTV which already has appropriate legislation: Deployed coverage infrastructure can be limited to specific areas, both indoors and outdoors and always inform the user about it, etc. However, there are also methods to preserve anonymity with RFID technology [53,79,80] which can be implemented in open city environments. In addition, the system may reduce the acquisition of unnecessary information which might involve intrusion into user privacy.

This technology provides installation options with passive or active receivers according to the desired range and complexity of the receiver. In the case of passive receivers, no power supply is required for transmitters [76,77], which reduces the drawbacks of consumption and portability for the user and enables its use for extended periods of time.

Regarding the economic cost of the proposal, the utilization of existing tags among the population reduces the cost of deployment. In this aspect, the physical nature of communication technology enables the use of metal urban furniture items as card reader antennas [38]. In this way, elements such as street lights, billboards, traffic signs or traffic lights can all operate as antennae. In addition, most of them are connected to the network, and then they can reduce the cost of the necessary elements for the information transmission towards the central system.

Finally, in terms of the ability for concurrent reading, the selected technology is able to work with a high level of traffic to simultaneously capture the location of a large number of tags located in the antenna reading radius. For example, the standard reader used in the experiments has a throughput up to 400 concurrent readings per second [81].

3.1.2. Location Infrastructure Design

The RFID technology mentioned in the previous section consists of two types of elements: The antennae or readers that emit the signals and receivers or tags. As mentioned, in the passive mode of this technology, receivers do not require power and can be embedded in objects familiar to users such as keychains, bracelets, keys or plastic cards worn by users.

Although this technology was conceived for identification processes, nowadays, it is also used for location tasks. Concerning this topic, research works are focused on the design of reader networks where each of them has associated geographical coordinates. Thus, from the received readings and deployed topology, it is possible to determine the position of the users (location).

The aim of this process is to design a network of RFID sensors deployed throughout the urban area of interest for reading the locations of citizens (e.g., shops, streets, avenues, parks, government buildings, etc.). The advantages associated with this technique make it easy to adapt to heterogeneous contexts (flexibility), and both indoor and outdoor environments (transparency). Moreover, adding new readers to the network is feasible according needs of the urban area management (scalability). The next figure (Figure 4) schematically shows the A-part of the overall architecture such a network of RFID sensors deployment in a city environment.

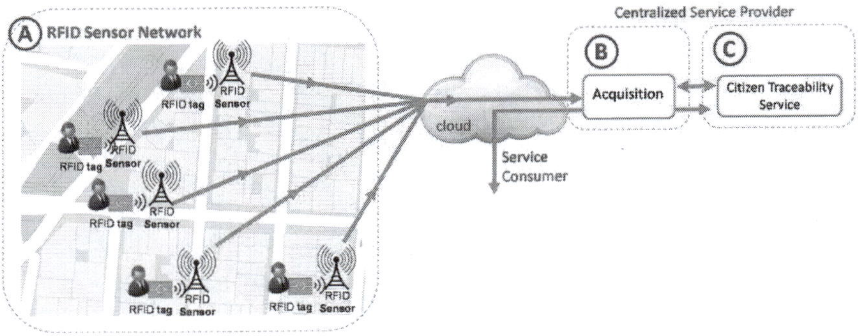

Figure 4. RFID sensor network deployment.

According to the features of RFID technology, it is possible to configure readers with coverage radii of several meters [76,77], so each of these readers can cover the sidewalk of a street entirely, as shown in Figure 5. The purpose is that collaboration of citizens for the activation of tags is not required and all they have to do is allow that they be worn on their clothes, carried in a pocket, bag or wallet in a transparent manner and interact as they move without the user being aware of it. Thereby, the tags worn by citizens are activated when they walk down the street and pass by one of the installed antennas.

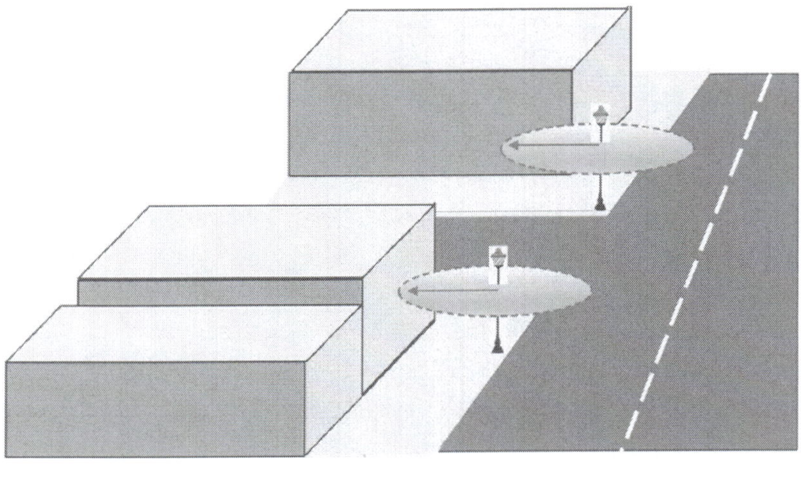

Figure 5. Installation and sensor coverage example.

The grid topology must be dense enough to cover the circulation in areas of interest. For practical reasons, tags used must be in some application that is already widespread and accepted by the citizens, while as mentioned, the new antennas can be complemented by already installed infrastructure.

With regards to the deployment of this additional infrastructure various criteria may be taken into account. For example, for practical criteria it may be decided to cover only areas of special interest in the city (downtown, tourist areas, etc.) or for economic criteria it may be decided to limit the grid to some areas of the city. In this regard, there may be options for installation of mobile antennae that can move from one part to another of the city when the area is sufficiently studied.

3.2. Method for Communication and Structuring of Citizens' Location

In the analyzed works based on RFID technology, the proposed solutions are focused on restricted areas, such as supermarkets, homes, hospitals, logistics centers, places of entertainment, etc. where the information transmission is performed directly through private intranets (not on public networks like the Internet) without reliability capabilities. In these proposals only an RFID tag type is used and processing needs have not been considered. However, after this analysis, it was identified that one of the challenges of the proposed system was the heterogeneity and lack of environmental control where it is mandatory to act in order to acquire the citizens' location information to build the movement flows. In this sense, any space in the city can be improved, without knowing the number or the type of tags to read or the capacity limits of the required processing. In this new scenario, it is necessary to propose a system with the technology and infrastructure indicated in the previous section, capable of distributing the logic of the system to ensure the workload processing, transmission and storage and structuring of information can be carried. This will establish the citizen flow and, therefore, the objectives will be addressed as outlined in the new scenario.

This is the objective of the task described in this section. More specifically, the goals which these process should cover are: To be capable of acquiring and processing the information independently of environment or data structure of the RFID tags (transparency), regardless of the number of readings performed by the park of sensors deployed in the network (scalability); to ensure the delivery and reception of each location to generate the different citizen flows (reliability) without citizen identification data (privacy); to ensure that the information is accessible from any place and on any device for the service consumers (availability and versatility); to offer complete information about the citizen movements, despite the lack of certain individual locations (completeness); to manage the systems resources in an optimal way (efficiency) and to distribute the load (performance), due to the high amount of location information that every reader will be able to receive and send to the central system; and, finally, to be interoperable in order to obtain information from different location formats or methods (interoperability), as the system is intended to take advantage of the already deployed elements and incorporate them for users who could might work with different protocols or formats.

In this sense, an analysis of techniques, patterns and design principles that guide and control the transformation process to achieve the proposed system and reach the mentioned goals has been necessary. The result of the process is a distributed architectural model based on the use of RFID Smart Sensors networks whose functionality will solve the problems (A-part of overall architecture), and a centralized cloud-based system which complements the information acquisition features. The RFID Smart Sensors basically consist of an RFID reader with computing and communication capabilities. The centralized cloud-based system will focus on the generation (B-part of overall architecture) and provisioning of citizens' flow (C-part of overall architecture) in an open and transparent way. The result of this process presents a reference model based on the integration of patterns and software design techniques. The novelty of this method lies in its combination and slight modifications to address the underlying problem and to solve existing deficiencies in current systems. This combination satisfies the use of RFID technology in uncontrolled and heterogeneous environments for determining the flow of citizens' movement in order to provide an efficient management of the resources of the cities. Figure 6 describes the processes involved in it, according to the methodology used with its main goals.

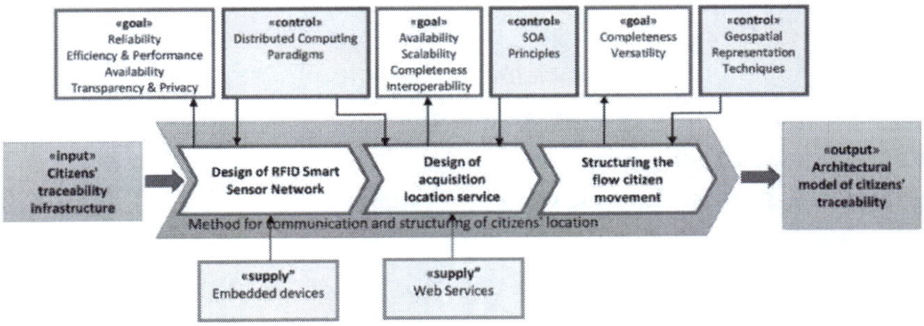

Figure 6. Modelling with Eriksson-Penker notation of the details of the process Design of the method for communication and structuring of citizens' location.

Next, we will describe the specifications of these elements in order to achieve the information acquisition, structuring, and communication goals.

3.2.1. Design of RFID Smart Sensor Network

Although there are related works which propose distributed architectural models based on sensors network to get the individual locations of the citizens (for example [42,45,48,50,53,56,58], these approaches do not cover necessary aspects (reliability, privacy, performance, efficiency, availability and transparency) to give a solution to the problem described. One of the main contributions and novelties of this paper is the provision of intelligence to sensors. This device is mainly responsible for reading the locations and ensuring the delivery of such information. So that, in order to provide the right route generation from each movement, it is necessary to design the infrastructure as a network of RFID Smart Sensors which include the necessary rationality to track citizens. Therefore, each Smart Sensor consists of an RFID reader, together with an element of embedded processing that provides computing and communication capabilities to support the necessary functionalities to reach established goals. These functionalities are supported by eight modules such as is shown in the schematic architecture of the RFID Smart Sensor (Figure 7).

One of the main functionalities included in the RFID Smart sensors is the capacity to receive information from different RFID tags types, transparently. About this issue, only the proposal presented in [45] proposes a solution that allows the capture of the location information from various types of RFID tags through the use of ontologies. Although this proposal would require a prior knowledge of the structure of information stored in the RFID tag to be read, an aspect that could mean that the management of ontology is untenable. It should be necessary to know all the existing tags types or every time is detected a new one to introduce the information into the ontology.

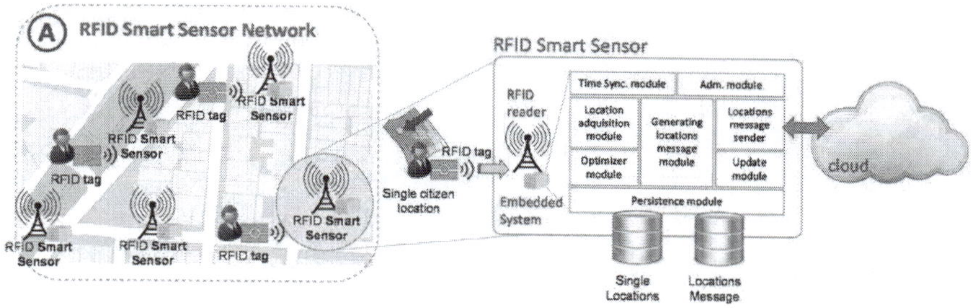

Figure 7. RFID smart sensor architecture.

The remaining works analysed can only capture tag information provided by themselves. Since it is required in the problem proposed, our solution is not dependent on the RFID tag. In the Smart Sensor proposed, it has included functionality to obtain, transparently, full information as mentioned, from the generation of a code from the information that has been read to identify the location of form unique. When a new acquisition of the same tag arrives, the same code would allow traceability is created. Although a security system is included to increase the privacy of citizens that is every day changes the code generation.

There is no wish to know if it is the same person carrying the tag, but the traceability of all citizens. These functionalities are performed by the Location acquisition module of the Smart Sensor described in Figure 7.

To perform reliability issue, we will implement the integration pattern known as guaranteed delivery [82]. This pattern determines an architectural design which prevents the loss of information messages transferred among the computer nodes that make up a distributed system. The pattern is based on the previous persistent message storage from a node in local disk. Until the destination node does not confirm message reception and storage, the source node will not delete the message from its local storage system. If the destination node never confirms message reception, the source node would retry sending it. Only with the use of the mentioned pattern, the complete delivery of the data cannot be ensured. The data could be sent and, because of the existence of communication problems, message reception confirmation may not arrive at the RFID station. In this case, if the message is not transmitted, the data would not be deleted and there would be a loss of information; or if the message were resent, data would be duplicated. To solve this issue, we focus on the correlation identifier pattern [82]. Each submission message will be marked with an identifier formed by a time stamp, which can establish the correlation for further processing, and by the identifier of the source RFID reader. Once communication is restored, the message would be sent again, being the acquisition module in the server the responsible for verifying information duplication. This way, we can ensure the complete reception of the necessary information to infer citizen flow patterns without losing or duplicating messages. This is another novelty functionality achieved through of patterns integration included in the RFID smart sensors. These functionalities and patterns are implemented by persistent module, location message sender and update module of the Smart Sensor shown in Figure 7.

Another key aspect to ensure delivery reliability in a distributed system, and it is not contemplated by the rest of the research works that provide RFID sensors distributed architectures, is the time synchronization of the system components. Most of the analysed proposals are focused on the tracking process [45,46,47,48,49,50,51,52,53,54,55,56,57,58]. The proposals focused on traceability process do not contemplate this question [42]. However, this aspect is necessary to achieve the tracing process. In order to generate reliable routes to manage resources in a better way, it is necessary to know the specific moment of the citizen location. A device provided with a badly synchronized clock would be sending the location at a wrong moment, the route information would not be valid, and it could even lead to wrong decisions. There are many algorithms that solve the problem of time synchronization, and choosing a particular one depends on the required accuracy of the system, the resources to be used, and the information that we want to obtain. For the present proposal, we will reject those systems that work with relative times and logical clocks, because, in our case, it is essential to synchronize all the elements of the system with a UTC reference system.

Depending on the type of network used and the precision required, we will have a great number of methods, among which are classical solutions such as Network Time Protocol (NTP), Simple Network Time Protocol (SNTP), and Time Transmit Protocol (TTP), or more advanced solutions such as Reference Broadcast Synchronization (RBS), Time Protocol on Sensor Networks (TPSN), and Flooding Time Synchronization Protocol (FTSP) [83,84].

In our case, the abovementioned patterns, together with time synchronization, ensure the reliability in the data delivery process. Also, the precision of the synchronization may perfectly tolerate a few-second lag, as the route would be relative to the local level and a range of several RFID readers would not appreciate such a lag. Moreover, once the server receives the location information from several neighbouring readers, there could be a data optimization and standardization regarding time. In the RFID Smart Sensor, this functionality is performed by the time synchronization module. This module is an autonomous and independent component that is executing periodically in background to update the clock system (Figure 7). In addition, the design proposed includes the following considerations in order to reduce the communication and management costs of the RFID smart sensor: Firstly, it is necessary to achieve an optimal transmission rate to take advantage of the bandwidth, which will be determined by the communication protocol established in the sensor. If we send the information via HTTP, the submitted location information could be less than the message wrapper of the communication protocol used. This circumstance would trigger penalties in each message sent. In our proposal, the number of locations in each sent message would be configurable in the Smart Sensor in order to achieve the best forwarding rate depending on the transport protocol used and the available bandwidth. The RFID Smart Sensor includes an administration module to manage and parameterized remotely the network sensors; secondly, the storage capacity of the Smart Sensors is limited, so it is essential to minimize local storage. In this sense, there would be a small modification of the guaranteed delivery pattern, in which not only would the message be erased to get the reception confirmation, but also the location data that were included in the message. With this design, once the RFID smart sensor obtains the location of a citizen, it stores it locally, creates a message from the citizen location data which is registered in the database, and stores them again. When the message is sent to the information system of citizen flow and is stored, the system returns reception confirmation to the client and the message is finally erased from the local system. This process is performed by the update module of the RFID smart sensor.

The remaining of the research works performs the communication process of the location data in controlled environments regardless of the communication reliability. For this reason, they do not propose the persistent capability at the sensor level and they have not contemplated the resources optimization of the sensors as our proposal.

Below, the proposed design works according to the following procedure: as the RFID reader is reading different citizen tags, these are received by the local location acquisition device through the acquisition module that performs the role of Coordinator.

Whenever an RFID tag content is received, the timestamp (only year/month/day) is attached to it and then the hash function is applied. The result of this action will be used to perform anonymity to the citizen. Then, this information is sent to the optimizer module, which is responsible for indicating if a citizen location must be stored or discarded depending if it is considered redundant data or not. If the information is valid, the acquisition module stores it in the local location database (persistent module) and informs the generating location message module, which uses the data to control if it reaches the optimal number of acquired locations to create a message that will be stored in the location message database. When the location message sender detects that there are new messages, it begins to transmit them by connecting to the citizen flow generation service or acquisition location service. The message will send the data through the available transport protocol.

At this moment, the RFID smart sensor will wait to the acknowledgement message location from citizen flow generation service. When the sensor receives it, then its update module is invoked, deleting the corresponding message and their individual locations from the database in the same transaction. This operation allows optimizing resources. If during a time previously configured the message acknowledgement was not received, the location message sender would return to try to send the message, as many times as was configured the device parameters. Before, the correct operation of network communications has to be checked.

3.2.2. Design of Acquisition Location Service

This service must provide an interface for data acquisition, which allows receiving location messages from users of smart sensors ensuring scalability and availability of the system, the reliability in data delivery and the interoperability with different types of sensors. This service corresponds with B-part of the overall architecture. The following figure shows the internal details of this module in conjunction with the smart sensor.

Scalability will be determined by two factors: firstly, the number of citizens who could interact with the system could grow indefinitely and, therefore, the data size to be acquired by the system could also grow in a reliable way. Secondly, the number of sensors deployed for the acquisition of citizen information may be high and could also grow considerably in order to obtain more data and to ensure accuracy in the flow determination. These two factors could make the citizen flow service provider become a bottleneck when it comes to receiving and processing the information. In order to avoid these situations, we propose two solutions that have been implemented: The first solution would be to separate the process of receiving and storing location messages, which requires less processing, from the transformation and processing of location data included in the messages. This solution would be provided by the message oriented middleware (MOM) paradigm or the software asynchronous queuing design pattern [85] (MOM component in Figure 8), establishing a point-to-point model based on message queues as an interface to the location acquisition service. This is a distributed computing model, which is asynchronous and transactional, and which acts as an intermediary message in the communication between a sender, the RFID Smart Sensor, and a consumer, the citizen flow service provider. In addition, this pattern enables the messages to remain, allowing us to implement the abovementioned guaranteed delivery pattern to ensure the reliability of the message communication; the second solution, which is complementary to the previous one, consists of the introduction of clustering techniques for load balancing using multiple nodes to be distributed and placing a messaging system for data acquisition in each of these nodes. Thus, the burden of data acquisition would be fairly distributed, and even the implementation of new acquisition nodes would be immediate.

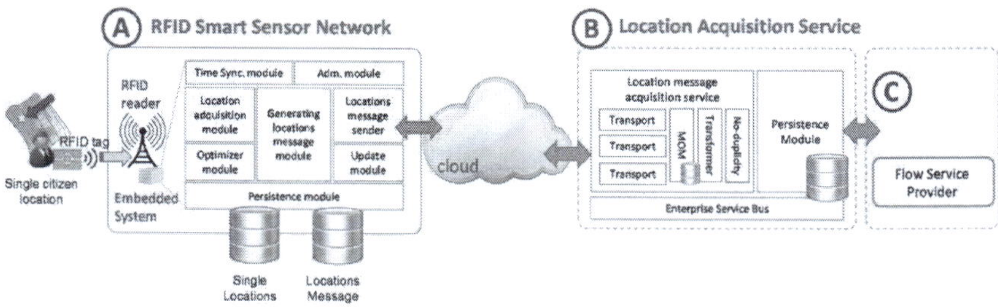

Figure 8. Location acquisition service architecture.

Although the use of MOM paradigm has been proposed in [45] to provide scalability in indoor tracking system processing, in this proposal the communication reliability between sensors and central system and fault tolerance of the central system are not contemplated. Moreover, no aspect of high availability and load balancing is contemplated on this functionality as the needs that require scalability and processing are limited by the application environment.

In relation to the availability, a centralized system collapse or failure may involve the loss of location data and, therefore, reduces the validity of the generation of citizen flow patterns. To solve this problem, clustering techniques could be added to the acquisition module in order to achieve high availability. This way, if one of the nodes falls, another one from the cluster would replace it.

Finally, another important issue to be considered when location data are sent from RFID smart sensors is the heterogeneity of the existing client devices in the market and the communication protocols they use to transmit data. There are RFID readers which have computer systems that allow communication through standards which are known and widely extended. However, others only include serial communication and a small low-cost embedded device should be included to transmit information. In addition, although the proposal mainly relies on collecting and acquiring citizen locations through RFID technology, the system could consider acquiring data through other types of location and positioning systems such as GPS or legacy data systems that provide location information. In turn, this would lead a diversity of message formats and data models.

In those cases, we have considered creating an infrastructure based on protocol bridging patterns, data model transformation and data format transformation [85]. The protocol bridging design pattern is implemented by different transports modules (Figure 8), although, by default, the transport module of the system described in this section is based on Rest style following the SOA paradigm. The Rest service contract has been defined by means of Rest APIs Modelling Language, RAML (Figure 9). The other two patterns are implemented by the transformer component shown in Figure 8. The goal of the transformers is to unify the message in order to be the performance of citizen traceability service provider. In this way, we can dynamically solve the heterogeneity problems of the message structure, the data format and the communication protocol used for respectively sending the acquisition module.

```
#%RAML 0.8
---
title: Citizens Location Service
baseUri: http://www.dtic.ua.es/v1.0
version: v1.0

/citizenlocations:
  post:
    description: create locations read from RFID smart sensors
    protocols: [HTTPS]
    body:
      application/json:
        example: |
          {
            "idrfid":1,
            "location":[
                 "lat":38.384993156837425,
                 "lng":-0.5133978999999727
            ],
            "locations":[
                 "citizen":["cid": 101010101,"pw": 64,"ts":1443723690],
                 "citizen":["cid": 101010101,"pw": 62,"ts":1443723695],
                 "citizen":["cid": 203330107,"pw": 62,"ts":1443723695]
            ]
          }
    responses:
      201:
        description: Locations have been successfully created.
        body:
          application/json:
            example: |
              {
                "messeage": "Locations have been successfully created."
              }
      400:
        description: Locations have not been created.
        body:
          application/json:
            example: |
              {
                "message": "Locations have not been created."
              }
```

Figure 9. RAML contract of location message acquisition service.

When the message is received through the correct transport module, the message is stored in a persistent management system (point to point queue messaging model based on the pattern asynchronous queuing [85]). Messages are stored waiting to be consumed and processed. From this moment, the system sends an acknowledgement message to the RFID smart sensor to confirm location message reception.

It could happen that the message had been received by the system, but the acknowledgement response sent by the system fails because the communication was interrupted. In this case, a message previously processed could be processed another time. This is the function of the no-duplicity module, to filter the messages and to avoid their reprocessing using the message identifiers (message_id and rfid_reader_id). When a message is received from the queue of the message system (MOM), their structure and data format is checked. Depending on this information could be possible the use of transformers (data model and format transformation patterns) [85].

All modules of the location acquisition module are supported by Enterprise Service Bus (ESB) platform. ESB is a powerful infrastructure for software integration and communication that enables the interoperability between the described modules.

3.2.3. Structuring the Citizen Movement Flow

From the requirements analysis performed in the preceding paragraphs, in this process we must identify the information required to ensure the generation of reliable citizen flows which will allow making decisions in the management of the city resources. In order to do so, the main goals are the provision of complete information concerning the geographical area under study (completeness), with the possibility of segmenting the information depending on the time of day, the day of the week and the time of year (versatility).

The most basic and essential entity in order to determine the citizen flow of a citizen, will be the location (L). This entity will be defined by the following tuple:

$$\mathrm{L} \equiv \{rfss\,(lo, la), cid, pw, ts\}$$

where:

- rfss (lo, la) represents the location of the RFID Smart Sensor determined by its longitude and latitude, which will allow us to know the position of the citizen on the map.
- cid represents the radio-frequency identification which is carried by the citizens and which allows us to identify them in a unique and anonymous way.
- pw it is the power of a certain card reading signal which will allow us to know the proximity to the RFID Smart Sensor antenna. Based on this, the information on the RFID Smart Sensor will be preprocessed in order to eliminate redundant locations with no added value which belong to the same user and will help determine possible directions of a citizen or trajectory changes with the data from other neighbouring RFID Smart sensor.
- ts it is the moment in time in which the RFID tag reading took place.

This data will be sent to the centralized system by the Smart Sensor, following the mechanisms described above to ensure delivery. The other data entity needed to determine the citizen flow is the deployed RFID smart sensor network itself. This information, together with the individual locations, will be completed to determine the route taken by a citizen. The sensor network entity will be determined by the following elements:

$$RFSSN \equiv \{RFSS_1, RFSS_2, \ldots, RFSS_n\}$$

$$RFSS_k \equiv \{id_k, lo_k, la_k, N_k\}$$

$$N_k \equiv \{RFSS_i \ldots RFSS_j\} \subset RFSSN$$

where:

- RFSSN represents the set of RFID Smart Sensors that make up the sensor network.
- RFSSk is a specific sensor defined by its unique identifier (idk), its position, determined by the longitude (lok) and latitude (lak), and the

set of its neighbouring smart sensors (Nk), which in turn will be a subset of the network (note that the geographical-neighbours of a sensor could not match with its real neighbours due to insurmountable obstacles or habits of movement of citizens).

The data received by the deployed sensor network should be structured and completed in order to build the citizen movement routes in the geographic area under study. Therefore, knowing the neighbouring sensors allows addressing potential locations losses on the route followed by a citizen and helps define the interpolation and completeness strategies according to methods known to do so.

The combination of the data collected will be used to generate on-demand citizen flows for specific criteria, such as time interval or geographical area, indicating their density or the possibility to foresee citizen movement flows according to behaviour patterns based on historical movement data.

3.3. Method for Provisioning Citizen Flow

The orientation of this research to the study of public spaces in the city has led us to design methods to provide the information generated openly and as a value added service. In this section existing technological paradigms are used to enable easy consumption by third parties. The next figure (Figure 10) depicts schematically this service (C-part of overall architecture) in conjunction with the rest of the centralized architecture.

According to the foreseen, the goals focus, firstly, in the on-demand supply of the citizen movement flows based on the needs and search criteria for consumers (segmentation). Secondly, it is important that the value-added service can be used by external applications (interoperability). Therefore, this task has been divided in two stages: The design of data search filters and the supply of the system as an interoperable service.

In the first stage, several filters have been established to search based on the consumer needs. The first filter would be aimed at determining the action area where we want to make the decision. This action area could be defined by different indicators with a greater or lesser degree of accuracy such as a specific location, a specific location and a radius of action, a specific route, an area previously defined by the urban policy makers, or, finally, some irregular area determined by the service user. The second filter would determine the time period about which we want to obtain citizen flow data for the specific area and which would be given by parameters that represent different precision levels, from a short period of time in the same day to a long period in the long term, which could be months or years. Another filter would be based on the density of routes, allowing the information about specific routes that exceed or are below a threshold and that can contribute to the modification of the use of city resources. For example, a route with low density for a long period of time may lead to the modification of a public transport route.

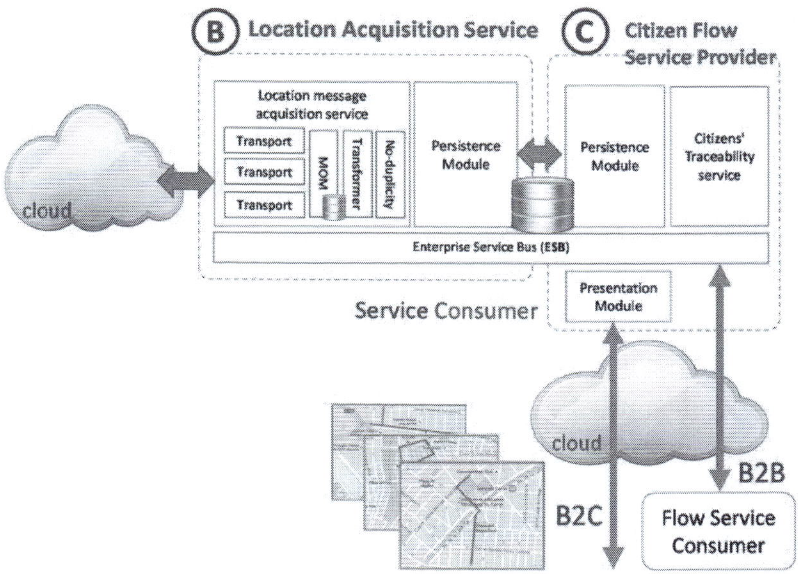

Figure 10. Location acquisition service architecture.

```
#%RAML 0.8
---
title: Citizen Flow Service
baseUri: http://www.dtic.ua.es/v1.0
version: v1.0

/citizensflows:
  uriParameters:
    zone:
    timeInterval:
    density:
  get:
    description: get citizens flows
    protocols: [HTTPS]
    responses:
      200:
        description: Get citizens flows.
        body:
          application/json:
            example: |
              {
                "flows":[
                   "flow":[
                      "density":20,
                      "color":"red",
                      "locations":[
                        "location":["lat":38.384993156837425, "lng":-0.5133978999999727],
                        "location":["lat":38.3850402521112, "lng":-0.5150179542449673],
                        "location":["lat":38.38557848104228, "lng":-0.5152110732940396]
                      ],
                   "flow":[
                      "density":10,
                      "color":"blue",
                      "locations":[
                        "location":["lat":38.38687357799348, "lng":-0.5135802902129849],
                        "location":["lat":38.38551120264542, "lng":-0.5130653060821255]
                      ]
                   ]
                ]
```

Figure 11. Citizen flow service RAML contract.

As mentioned before, the second stage focuses on securing the interoperability of the service with the consumers. In order to do so, this service should offer the following two models: A first approach based on Business to Business (B2B) paradigm will allow connecting the consumer system with the service to incorporate it as another one of its organizational processes, but improving the current system by adding more decision data. Both cases will be based on the service supply according to the principles of the Service Oriented Architecture (SOA) paradigm, [85] implementing through Rest service technology and offering the resulting information using a format that is structured as JSON, so that it is easily treated. The service contract, modelling with RAML, is shown below (Figure 11). As can be seen, the service contract includes the filters defined previously to carry out searches by means of next parameters: Action zone, time period and flow density; a second approach, based on the Business to Customer (B2C) paradigm, oriented to offer a human-machine interface which will allow a user to interact directly with the system, for example through a thin client such as a Web browser. The access interface of this module defines the search filters described previously. When the Rest service receives these parameters, this one returns the citizens flows and their density drawn on cartographic map. In both cases, the citizen' flows will be generated from the information collected by RFID Smart Sensors and the information of the sensors topology grid dynamically and on demand, based on the input parameters representing the filters entered by the consumer. The generation of flows combined through individual routes of the same or multiple users may be caused by adding readings for the specific filters through a clustering process.

The calculation of these added routes can be performed through several procedures by applying a similarity function to them. For example, some of the functions of this type used are the comparison of the common points of the sequence, and the comparison of distances or times used in them. These procedures are beyond the scope that defines this architecture, although they are being considered in the line of work developed by the research group.

4. IMPLEMENTATION AND VALIDATION

4.1. Implementation

In order to be able to validate the proposal, a prototype of the architectural model shown previously has been implemented. The RFID Smart Sensor prototype is shown in Figure 12. The sensor used is marketed by RFID Controls Enterprise (Murcia, Spain). This is a highly configurable and flexible sensor with two antennas [81]. It has a maximum reading rate of 400 tags/s and its maximum reading potency is 1 W. This device even allows connection to two antennas through two 50 Ohm MCX connectors with 6.5 dB gain, providing circular polarization and a radiation pattern with 60°/60° beamwidth (Figure 12a).

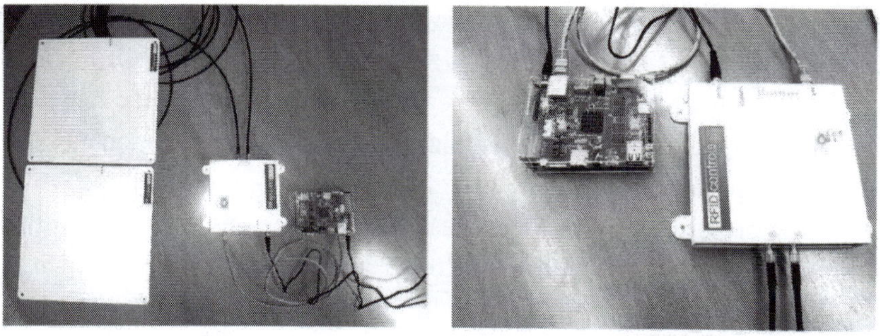

(a) (b)

Figure 12. RFID Smart Sensor prototype: (a) RFID Smart Sensor with antennas; (b) RFID reader with an embedded system.

In order to provide the sensor with the necessary intelligence for implementing the aforementioned modules, an embedded device based on an ARM processor has been used. This device has a low energy consumption and it has been integrated with the RFID reader through its Ethernet interface. Specifically, the Cubieboard3 cortex-A7 dual-core 2 Gb embedded device has been used, provided by Cubietruck Enterprises [86] (Figure 12).

The RFID Smart Sensor modules designed in the architectural model have been implemented using C language. Reading results are saved in CSV files and stored in a system directory. To implement the clock synchronization module the NTP protocol has been used, because is a reliable protocol with suitable error rates for our system. For this reason, an NFS server has been enabled, which is used to update the system clock by means of the synchronization module implemented in C language.

The proposed architecture of the citizen track and trace system is based on Enterprise Service Bus integration infrastructure. This solution is highly scalable and the software design patterns specified in the architectural model have been implemented over this platform. Specifically, the open source Mule ESB has been used, implemented two Mule ESB services flows to implement the Location Message Acquisition Service and the Flow Generation Service, shown in Figure 13a,b, respectively.

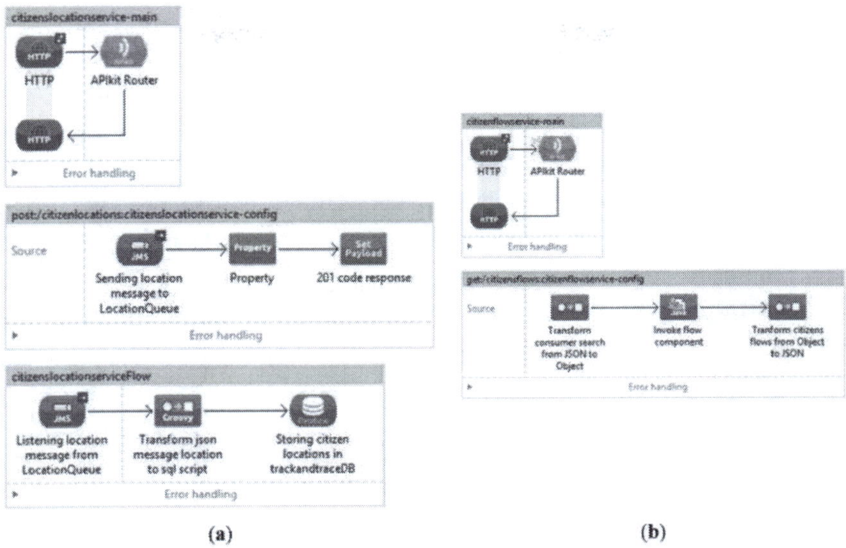

Figure 13. Mule flow. (a) Location message acquisition service; (b) Citizens flows generation service.

In order to provide data persistence, the MySQL community version database manager has been used. The Search Module for flow generation has been implemented using Java Language. This module is embedded and invokes the Mule service flow. Finally, the B2C Presentation Module has been developed through Web technologies like CSS, HTML5 and JavaScript using the AngularJS framework, in order to provide more accessibility and usability to the final users.

In order to achieve better analysis and interpretation of the experimentation results obtained in the case study described below, we have previously made a technical testing of the smart sensor and the RFID technology. The aim is to know beforehand their capabilities and limitations to certain situations that may occur.

First, we proved that with the power supplied by the sensor (1 W) and with the well-focused antenna the reader is capable of reading tags up to 10 meters away. Secondly, reading tests were performed simulating actual portability by citizens, i.e., placing tags hidden in clothes, pockets, purses, handbags or backpacks. These experiments significantly reduce the sensor range down to 3 m, but a good reading rate is still maintained in these cases.

The most serious problem encountered for our purpose is that the human body is opaque to the RF signal, so that the tag will be hidden from the sensor when there is a human obstacle. This case can be common in urban settings with a high density of pedestrians. For example, in Figure 14a, the two persons on the left are in a shadow area produced by another person. However, this problem can be solved by distributing the antennas of the smart sensor in a configuration that minimizes occlusions and signal shadow areas. Following the previous example, Figure 14b shows a configuration that reduces the space not visible to the sensor.

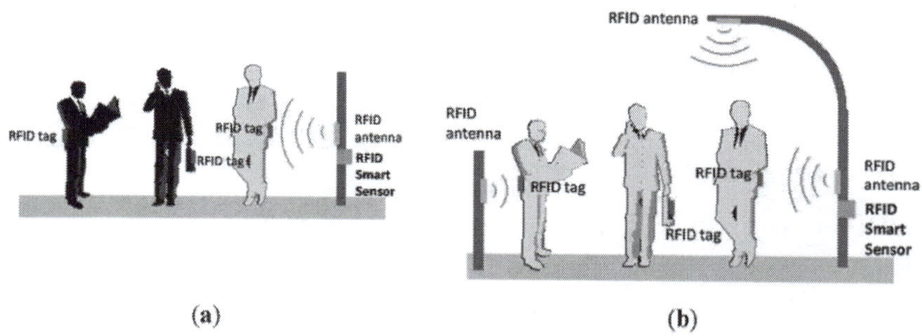

Figure 14. Configuration of the antennas and coverage of RFID sensors: (a) One antenna on the right side of a sidewalk; (b) Three antennas on the right, left and top of a sidewalk.

Regarding the other capabilities of the Smart sensor, the experiments proved the ability to synchronize with other sensors, the combination of multiple readings from the same person and the capability of sending information without loss. Through log systems of the proposal of the traceability of reading, we analyse sent and received citizens' locations and we verify that all locations readings are stored in the RFID Smart Sensors Network and they were received by Location Acquisition Service. In this way, the reliability of the proposed architecture is validated.

4.2. Case Study

In order to validate our proposal as well as to study a practical implementation of the concepts treated, a simulated use case in a real environment for tracking and tracing is proposed in this subsection. In our example, we consider the campus of the University of Alicante as a test scenario for the deployment of the proposed system. The case study will provide a better understanding of how various campus resources are used based on student and staff spatial movement and temporal behaviour. The ultimate goal of this case is to help the university improve management of the support services on campus.

The experiments performed will also result in a number of spatial and temporal visualizations of big data from these sources. The reason to carry out a project on a campus is the advantage of disposing of something similar to a little city, but more accessible. In a way, since it has different areas (rooms, labs, small shops, restaurants, open areas, etc.), diverse activities (academic, administrative, social, small health centres, sports…) the university campus can be seen as a small city. In addition, it has the facilities to make it as a perfect candidate to test our system under diverse conditions in order to find out citizen movement flow patterns. It is always far less complicated to locate antennas, RFID readers and the rest of technology within the university campus than throughout a city. In addition, it is possible to monitor the different states of every single location point, allowing testing of what are the best and most suitable places for the whole system. Therefore, in the matter of fact, we can see our system implemented in the framework of the campus as a prior step for a smart city which is much bigger, although it can be benefit from this previous stage.

The criterion for placement of the readers has been to monitor the most likely movements of the students (cafeteria-classrooms, classrooms-library, bus station-classrooms, etc.) as a preliminary step to a more dense deployment. Students were not informed that they were taking part in this experiment, only academic campus authorities were informed to request the appropriate permissions for the deployment of the readers. Due to this, no tags were distributed among individuals, but our intention was to first validate the transparency and anonymity of the readings, the ability to read tags and carried by people (on clothing, tickets, identification or transport cards, other RFID tags) without their cooperation and the capacities of transmission and processing of the architecture.

The experiment was conducted during one week. Through a visual monitoring of the antennas the number of students who passed through the sensor radius was counted. It was about 2K per day, with a tag read rate of about 5%. The amount of simultaneous readings of this data remained low, although, in most cases, when the tags were read by an antenna, they were read again by the next antenna on the same route.

From the study of these results, the reasons that lead the low rate of readings obtained can be: (1) the still low level of integration of RFID tags on clothing and other objects carried by users; (2) the lack of power in the readers to reach tags hidden in the wallet in the pocket of the user; and (3) the incompatibility of technologies with short-range tags (as NFC) to access ID cards or credit cards.

As to the first drawback (1), we think that will be resolved eventually, as the steadily increasing pace of implementation of RFID technology in everyday objects such as clothes or accessories will increase the reading rate. The second problem (2) can be overcome using directed antennas with a low angle of beamwidth coverage and high gain antennas rather than general purpose ones since in outdoor spaces, the signal can be dispersed more or noise sources can even appear [87,88]. In this regard, it is necessary to check regional regulations to find out the maximum allowable radiated power and antenna beamwidth in the UHF RFID bands. Finally, regarding the third cause (3), we think that this kind of RF technology belongs to a set of applications that should not be accessible, so to remain inaccessible, reinforcing anonymity of identity and user data.

In all cases, the readings collected and sent to the server correspond to a tag identifier on clothing or an accessory. This information consists, among others, of the following fields: An ID tag of different length, IP address of the reader, received signal power and DateTime (i.e., 30396062c3c324800009e6 77, 169.254.247.184, −74, 03/02/2015 08:55:59.121). Under no circumstances is personal user information collected. Due to the functionality included in the Location Acquisition Module which codifies RFID tags content through hash functions, we could not get any data of those RFID tags. In addition, any of the read RFID tags could be related themselves between readings obtained from different days.

The central system is hosted on a server on the campus intranet. The network traffic generated and the system processing times are trivial for a standard workstation, as the number of simultaneous readings remained low all the time, reaching traffic peaks of up to 3 reading/s.

To view the results we used a web-based user interface, which shows the positions of the readers and users-route through a third party application (Google Maps JavaScript API v3). For this case study, the next figure (Figure 15) shows a map of the University of Alicante showing the installation locations of the reader antennas and two temporal visualizations along the day where one can distinguish the main flow pattern per hour.

The results show that the movement flows of students change throughout the day. In Figure 15a it can be seen that the antenna located near the tram and bus platform receives students first, and from there, they are read by sensors located in other areas of the university as they move around the campus, both via inbound and outbound scenarios. In these hours of the morning, the movements are primarily targeted towards classroom buildings, offices and the library. Figure 15b shows the movements at other times of the day, when students walk to the campus exit and to restaurants and cafes to have lunch.

In this case study, the students' movements read by the system may seem obvious at first glance, since they coincide with the normal movement at that time of day. However, this case study provides the starting point for further deployment in which the movements of students and staff of the university can be analyzed in more detail. This information can help managers to provide better services and improve user satisfaction by providing resources where they are needed at any given moment.

In light of these results, we can assert that this experiment verifies the validity of the proposed hypothesis regarding the capabilities of RFID sensing technologies to read the positions of citizens outdoors without needing any kind of collaboration. The correction of distributed architecture to collect, synchronize and to process readings received while maintaining the anonymity of the users involved, it is also validated. The workload has not been a problem in this context due to the low rate of simultaneous readings. It is expected that a larger-scale deployment will allow us to validate the system in this regard. This result provides a promising perspective in the development of IoT for interaction with users.

Finally, a summary of the main objectives achieved by the proposed architecture in comparison with the capabilities of the other proposals is shown in Table 3. The comparison is focused on the results obtained in the experimentation, indicating what objectives would be achieved with each proposal. The table illustrates that related works offer partial solutions to complete all the stated objectives.

5. CONCLUSIONS

The design of smart cities uses the information generated in the city itself for decision making aimed at improving the quality of life of citizens. Infrastructure planning is one of the most pressing challenges of the development of cities and usually the handling of this issue does not only correspond to practical criteria, but also social, political, historical or other aspects. Knowledge of citizens' flow movements is very valuable information for designing transportation routes, as well as the optimal positioning of resources. With this knowledge, it is feasible to perform infrastructure designs which can minimize deviations from real movement flows. It is also possible to increase transport dynamic strategies in accordance with these needs to meet the movement habits of people by time of day, day of week or time of year to increase the use of infrastructure and thus improve the comfort of the citizens. The latest ICT advances can offer promising solutions to this specific problem by providing efficient and effective ways for handling large amounts of unstandardized and distributed information.

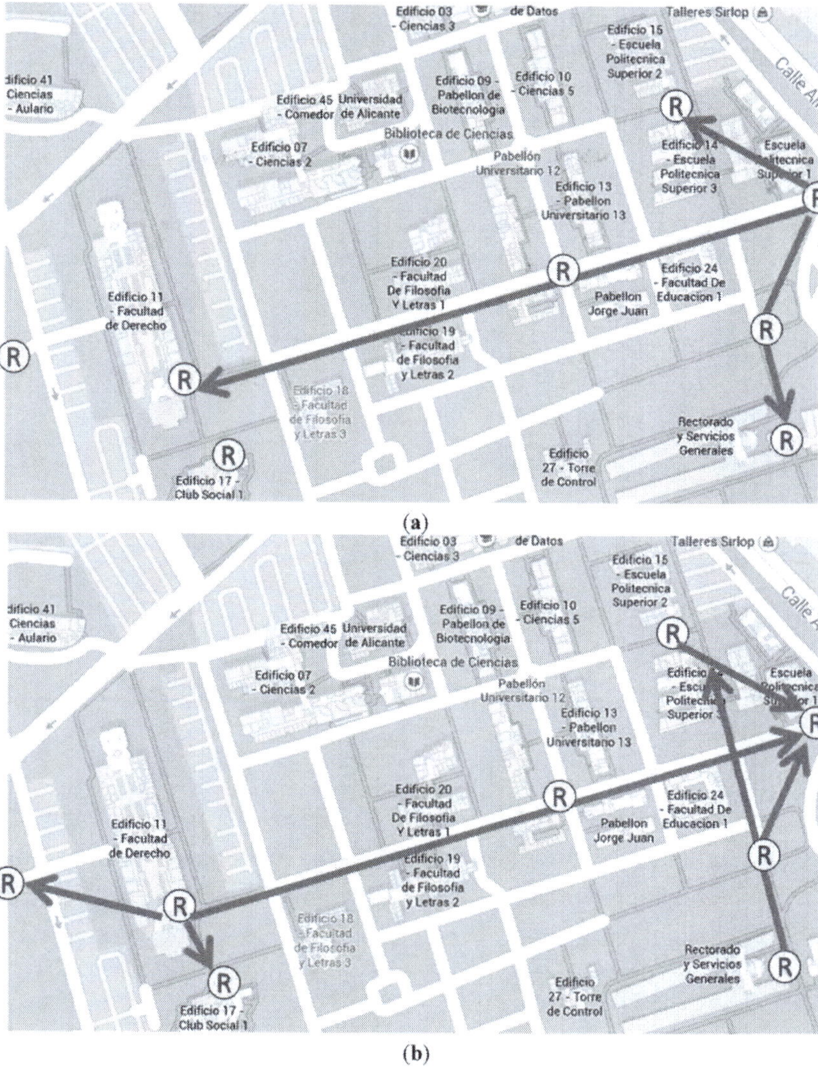

Figure 15. Temporal visualizations on the University Alicante campus with flow patterns: (a) temporal flow between 7.00 and 9.00 h; (b) temporal flow between 12.00 and 14.00 h

Table 3. Comparison of the obtained results between our proposal and the related works.

Research Work	Main Functionality	Transparency & Anonymity	Working Environments	Reliability	Energy Consumption	Scalability
Proposed architecture	Track & trace	Yes. No user interaction & Independency of RFID tags	Indoor & outdoor	Yes	No (passive tags)	Yes. (Distributed and decoupled approach)
GPS proposals [27–30]	Track & trace	No. User interaction	Outdoor	No	High	No
Wi-Fi proposals [16,34]	Track	No. User interaction	Indoor & outdoor	No	High	No
CCTV	Surveillance	No. Surveillance cameras	Indoor & outdoor	No	High	Yes
Science museum [42]	Track & trace	No. User/Object interaction & Dependency of own RFID tags	Indoor	No	No (passive tags)	No
Oyster Card [40]	Track & trace	No. Dependency of own RFID tags	Indoor & outdoor	No	No (passive tags)	No
Blind/inside RFID [47,53]	Track	No. User interaction & Dependency of own RFID tags	Indoor & outdoor	No	No (passive tags)	No
REACT [58]	Track	No. User/Object interaction & Dependency of own RFID tags	Outdoor	No	No (passive tags)	Yes
LANDMARC [41,46]	Track	No. User/Object interaction & Dependency of own RFID tags	Indoor	No	Yes (active tags)	No
Peer-to-Peer Networks [57]	Track	No. User/Object interaction & Dependency of own RFID tags	Indoor	No	Yes (active tags)	Yes
iWalker [44]	Track	No. User/Object interaction & Dependency of own RFID tags	Indoor	No	No (passive tags)	No
Other [45,48,50–52,54–56]	Track	No. User/Object interaction & Dependency of own RFID tags	Indoor	No	No (passive tags)	In controlled env. only [45,55]

The proposal described in this paper is the design of an architecture that can read, process and provide as a service the movements of individuals in a given geographic area based on real-time information acquired from the individuals themselves as they move along the city in an unassisted way to improve the resources management of cities.

The research presented in this paper includes an exhaustive analysis of the state of the art of techniques for solving the citizen traceability problem in cities. Its conclusions have allowed us to identify limitations of the current methods and technologies on the basis of the following main criteria: Heterogeneity, maintaining anonymity, no user intervention, uncontrol and cost. A comprehensive architectural model has been conceived using RFID wireless technologies with passive tags present in many devices, accessories or the clothes of the citizens to perform track and tracing tasks in city environments. As a novelty, we have designed an RFID smart sensor to read, process and to send the information of the users' locations in a scalable and reliable way to ensure the delivery and reception of each location. Finally, we have developed a computational architecture which integrates all the elements and modules implied to provide cloud services of citizens' traceability for B2B and B2C, consumers based on SOA principles.

The way of achieving of goals established and solution of problems raised have been the main contributions of this research. The solution has been in most cases to identify techniques and patterns that partially solve these issues to integrate them and offer a global solution. It is worth mentioning that the integration has brought changes in patterns and adjusting these elements to offer a comprehensive solution we have not found in the literature reviewed. In addition, these solutions have been encapsulated in functions and distributed in elements building the architecture to ensure the requirements of the problem. As a result, a network of intelligent sensors is provided, which is a novel solution for the current track and trace systems, fully coordinated and a central system is added to provide the citizens' flow.

The work involved in this research fits into the concept of Smart City aimed at improving lives of citizens using current technology of communications, cloud computing and smart sensing. In this regard, some challenges for future work arise, such as combining other communication technologies (GPS, Wi-Fi, etc.), integrating other sensing devices to improve monitoring (smart phones, etc.), studying mechanisms to optimize the network, new sensors deployed to cover the largest area possible or applying the presented architecture in a real city scenario. Moreover, the combination of these communication technologies with social media data such as Twitter and Facebook, will help develop broader patterns of citizen behaviour and compromise, service and facility utilization.

ACKNOWLEDGMENTS

We would like to express our acknowledgement to the RFID Controls Company for proving the necessary infrastructure to make possible the experimentation.

AUTHOR CONTRIBUTIONS

All authors were involved in the foundation items. All authors wrote the paper and read and approved the final manuscript.

.

REFERENCES

1. Neirotti, P.; de Marco, A.; Cagliano, A.C.; Mangano, G.; Scorrano, F. Current trends in Smart City initiatives: Some stylised facts. Cities 2014, 38, 25–36.
2. Smart Cities Council. Available online: http://smartcitiescouncil.com (accessed on 9 June 2015).
3. OECD, Governing the City. Available online: http://dx.dir.org/10.1787/9889264226500-en (accessed on 9 June 2015).
4. European Smart Cities. Available online: http://www.smart-cities.eu/ (accessed on 9 June 2015).
5. Fietkiewicz, K.; Pyka, S. Development of informational cities in Japan: A regional comparison. Int. J. Knowl. Soc. Res. 2014, 5.
6. Fazio, M.; Puliafito, A.; Villari, M. IoT4S: A new architecture to exploit sensing capabilities in smart cities. Int. J. Web Grid Serv. 2014, 10, 114–138.
7. Atzoria, L.; Ierab, A.; Morabito, G. The Internet of things: A survey. Comput. Netw. 2010, 54, 2787–2805.
8. Mitton, N.; Papavassiliou, S.; Puliafito, A.; Trivedi, K.S. Combining Cloud and sensors in a smart city environment. EURASIP J. Wirel. Commun. Netw. 2012, 2012, 247.
9. Stankovic, J.A. Research directions for the internet of things. IEEE Internet Things J. 2014, 1, 3–9.
10. Gubbia, J.; Buyyab, R.; Marusica, S.; Palaniswamia, M. Internet of things (IoT): A vision, architectural elements, and future directions. Future Gener. Comput. Syst. 2013, 29, 1645–1660.
11. Choi, S.H.; Yang, B.; Cheung, H.H.; Yang, Y.X. RFID tag data processing in manufacturing for track-and-trace anti-counterfeiting. Comput. Ind. 2015, 68, 148–161.

12. Shia, J.; Lia, Y.; Heb, W.; Sima, D. SecTTS: A secure track & trace system for RFID-enabled supply chains. Comput. Ind. 2012, 63, 574–585.
13. Kang, Y.S.; Lee, Y.H. Development of generic RFID traceability services. Comput. Ind. 2013, 64, 609–623.
14. Muñoz-Gea, J.P.; Malgosa-Sanahuja, J.; Manzanares-Lopez, P.; Sanchez-Aarnoutse, J.C. Implementation of traceability using a distributed RFID-based mechanism. Comput. Ind. 2010, 61, 480–496.
15. Todorovica, V.; Neagb, M.; Lazarevica, M. On the usage of RFID tags for tracking and monitoring of shipped perishable goods. Proced. Eng. 2014, 69, 1345–1349.
16. Isaak, J. Cell Phone WiFi Used to Track Your Location. Available online: http://ieeessit.org/2014/01/16/cell-phone-wifi-used-to-track-your-location-2/ (accessed on 9 June 2015).
17. International Telecommunication Union. Aggregate Data 2006–2013 ICT Data for the World, by Geographic Regions and by Level of Development; ITU: Geneva, Switzerland, 2014.
18. Kelly, J.; Jones, P.; Barta, F.; Hossinger, R.; Witte, A.; Wolf, A.C. Successful Transport Decision-Making: A Project Management and Stakeholder Engagement Handbook; European Commission: Brussels, Belgium, 2004.
19. Allport, R.; Brown, R.; Glaister, S.; Travers, T. Success and Failure in Urban Transport Infrastructure Projects; KPMG Intternational: Amsterdam, The Netherlands, 2008.
20. Bickerstaff, K.; Tolley, R.; Walker, G. Transport planning and participation: The rhetoric and realities of public involvement. J. Transp. Geogr. 2002, 10, 61–73.
21. Galland, S.; Lamotte, O.; Gaud, N. MetroB: Evaluation and simulation of public transport system. In Proceedings of the IET International Conference on Smart and Sustainable City, Shanghai, China, 6–8 July 2011.
22. Giannotti, F.; Nanni, M.; Pedreschi, D.; Pinelli, F.; Renso, C.; Rinzivillo, S.; Trasarti, R. Mobility data mining: Discovering movement patterns from trajectory data. In Proceedings of the Second International Workshop on Computational Transportation Science, San Jose, CA, USA, 2 November 2010; pp. 7–10.

23. Batty, M.; Axhausen, K.; Giannotti, F.; Pozdnoukhov, A.; Bazzani, A.; Wachowicz, M.; Ouzounis, G.; Portugali, Y. Smart cities of the future. Eur. Phys. J. Spec. Top. 2012, 214, 481–518.
24. Schmitt, G. A planning environment for the design of future cities. In Digital Urban Modelling and Simulation; Springer: Berlin, Germany, 2012; pp. 3–16.
25. Bajaj, R.; Ranaweera, S.L.; Agrawal, D.P. GPS: Location-tracking technology. Computer 2002, 35, 92–94.
26. Moloo, R.K.; Digumber, V.K. Low-cost mobile GPS tracking solution. In Proceedings of the 2011 International Conference on Business Computing and Global Informatization, Shanghai, China, 29–31 July 2011; pp. 516–519.
27. Trasarti, R.; Rinzivillo, S.; Pinelli, F.; Nanni, M.; Monreale, A.; Renso, C.; Pedreschi, D.; Giannotti, F. Exploring Real Mobility Data with M-atlas. In Machine Learning and Knowledge Discovery in Databases; Springer: Berlin, Germany, 2010; pp. 624–627. [Google Scholar]
28. Yuan, J.; Zheng, Y.; Zhang, C.; Xie, W.; Xie, X.; Sun, G.; Huang, Y. T-drive: Driving directions based on taxi trajectories. In Proceedings of the 18th ACM SIGSPATIAL International Conference on Advances in Geographic Information Systems, San Jose, CA, USA, 2–5 November 2010; pp. 99–108.
29. Trasarti, R.; Nanni, M.; Giannotti, F.; Pedreschi, D.; Renso, C. A query language for mobility data mining. Int. J. Data Warehous. Min. 2011, 7, 24–45.
30. Iqbal, M.U.; Lim, S. Privacy implications of automated GPS tracking and profiling. IEEE Technol. Soc. Mag. 2010, 29, 39–46.
31. Michaela, K.; Clarke, R. Location and tracking of mobile devices: Überveillance stalks the streets. Comput. Law Secur. Rev. 2013, 29, 216–228.
32. Clarke, R.; Wigan, M.R. You are where you've been: The privacy implications of location and tracking technologies. J. Locat. Based Serv. 2011, 5, 138–155.
33. Dahunsi, F.; Dwolatzky, B. An empirical investigation of the accuracy of location-based services in South Africa. J. Locat. Based Serv. 2012, 6, 22–34.
34. Zhang, Y.; Li, L.; Zhang, Y. Research and design of location tracking system used in underground mine based on WiFi

technology. In Proceedings of the International Forum on Computer Science-Technology and Applications, Chongqing, China, 25–27 December 2009.
35. Chen, P.-C. A cellular based mobile location tracking system. In Proceedings of the IEEE 49th Vehicular Technology Conference, Houston, TX, USA, 16–20 May 1999.
36. Mitra, S.; DasBit, S. On location tracking and load balancing in cellular mobile environment-a probabilistic approach. In Proceedings of the International Conference on Electrical and Computer Engineering, Dhaka, Bangladesh, 20–22 December 2008; pp. 121–126.
37. Chawla, V.; Ha, D.S. An overview of passive RFID. IEEE Commun. Mag. 2007, 45, 11–17.
38. Weinstein, R. RFID: A technical overview and its application to the enterprise. IT Prof. 2005, 7, 27–33. [Google Scholar] [CrossRef]
39. Ni, L.M.; Zhang, D.; Souryal, M.R. RFID-based localization and tracking technologies. IEEE Wirel. Commun. 2011, 18, 45–51.
40. Bagchi, M.; White, P.R. The potential of public transport smart card data. Transp. Policy 2005, 12, 464–474.
41. Ni, L.M.; Liu, Y.; Lau, Y.C.; Patil, A.P. LANDMARC: Indoor location sensing using active RFID. Wirel. Netw. 2004, 10, 701–710.
42. Kanda, T.; Shiomi, M.; Perrin, L.; Nomura, T.; Ishiguro, H.; Hagita, N. Analysis of people trajectories with ubiquitous sensors in a science museum. In Proceedings of the IEEE International Conference on Robotics and Automation, Roma, Italy, 10–14 April 2007; pp. 4846–4853.
43. Pandit, A.A.; Talreja, J.; Mundra, A.K. RFID tracking system for vehicles (RTSV). In Proceedings of the First International Conference on Computational Intelligence, Communication Systems and Networks, Indore, India, 23–25 July 2009.
44. Kulyukin, V.; Kutiyanawala, A.; LoPresti, E.; Matthews, J.; Simpson, R. iWalker: Toward a rollator-mounted wayfinding system for the elderly. In Proceedings of the IEEE International Conference on RFID, Las Vegas, NV, USA, 16–17 April 2008.
45. Li, Z.; Chu, C.-H.; Yao, W. SIP-RLTS: An RFID location tracking system based on SIP. In Proceedings of the IEEE International Conference on RFID, Las Vegas, NV, USA, 16–17 April 2008.

46. Han, K.; Cho, S.-H. Advanced LANDMARC with adaptive knearest algorithm for RFID location system. In Proceedings of the International Conference on Network Infrastructure and Digital Content, Beijing, China, 24–26 September 2010; pp. 595–598.
47. Willis, S.; Helal, S. A Passive RFID Information Grid for Location and Proximity Sensing for the Blind User; Technical Report Number TR04-009; University of Florida: Gainesville, FL, USA, 2004.
48. Iturralde, D. A new system based on web services and RFID for tracking people in a pervasive mining environment. In Proceedings of the Latin-America Conference on Communications (LATINCOM), Santiago, Chile, 24–26 November 2013.
49. Buccafurri, F.; Lax, G.; Nicolazzo, S.; Nocera, A. A privacy-preserving solution for tracking people in critical environments. In Proceedings of the 38th International Computer Software and Applications Conference Workshops (COMPSACW), Vasteras, Sweden, 21–25 July 2014; pp. 146–151.
50. Nohara, K.; Tajika, T.; Shiomi, M.; Kanda, T.; Ishiguro, H.; Hagita, N. Integrating passive RFID tag and person tracking for social interaction in daily life. In Proceedings of the IEEE 17th International Symposium on Robot and Human Interactive Communication, Munich, Geramny, 1–3 August 2008; pp. 545–552.
51. Germa, T.; Tajika, T.; Shiomi, M.; Kanda, T.; Ishiguro, H.; Hagita, N. Vision and RFID data fusion for tracking people in crowds by a mobile robot. Comput. Vis. Image Underst. 2010, 114, 641–651.
52. Cucchiara, R.; Fornaciari, M.; Haider, R.; Mandreoli, F.; Prati, A. Identification of intruders in groups of people using cameras and RFIDs. In Proceedings of the ACM/IEEE International Conference on Distributed Smart Cameras (ICDSC), Ghent, Belgium, 22–25 August 2011.
53. Foster, K.R.; Jaeger, J. RFID inside. IEEE Spectrum, 1 March 2007. Available online: http://spectrum.ieee.org/computing/hardware/rfid-inside (accessed on 10 June 2015). [Google Scholar]
54. Matic, A.; Osmani, V.; Mayora, O. RFID-based system for tracking people: Approaches to tagging demented patients. Lect. Notes Inst. Comput. Sci. Soc. Inform. Telecommun. Eng. 2011, 70, 60–65.
55. Xiong, Z. Hybrid WSN and RFID indoor positioning and tracking system. EURASIP J. Embed. Syst. 2013, 2013, 6.

56. Hsu, C.-C.; Chen, J.-H. A novel sensor-assisted RFID-based indoor tracking system for the elderly living alone. Sensors 2011, 11, 10094–10113.
57. Hui, F.C.P.; Chan, H.C.B.; Fung, S.H. RFID-based location tracking system using a peer-to-peer network architecture. In Proceedings of the International MultiConference of Engineers and Computer Scientists, Hong Kong, China, 12–14 March 2014.
58. Lin, X.; Lu, R.; Kwan, D.; Shen, X. REACT: An RFID-based privacy-preserving children tracking scheme for large amusement parks. Comput. Netw. 2010, 54, 2744–2755.
59. Steffen, R.; Preißinger, J.; Schöllermann, T.; Müller, A.; Schnabel, I. Near field communication (NFC) in an automotive environment. In Proceedings of the 2nd International Workshop on Near Field Communication, Monaco, 20 April 2010; pp. 15–20.
60. Firence Card. Available online: http://www.firenzecard.it/?lang=en (accessed on 9 June 2015).
61. ExxonMobil's Speed-Pass. Available online: http://www.speedpass.com (accessed on 9 June 2015).
62. Washington Metropolitan Area Transit Authority. Shuttle Services at Metro Facilities Report; WMATA: Washington, DC, USA, 2011.
63. Collins, J. Gift Cards Go Contactless. Available online: http://www.rfidjournal.com/articles/view?762 (accessed on 9 June 2015).
64. Charette, R. Wal-mart to track clothing with RFID Tags. IEEE Spectrum, 27 July 2010. Available online: http://spectrum.ieee.org/riskfactor/computing/it/walmart-to-track-clothing-with-rfid-tags (accessed on 9 June 2015).
65. Marks & Spencer Embraces Change. RFID J. 2014. Available online: https://www.rfidjournal.com/purchase-access?type= Article &id=11952&r=%2Farticles%2Fview%3F11952 (accessed on 9 June 2015).
66. Zara Builds Its Business around RFID. The Wall Street Journal, 16 September 2014. Available online: http://www.wsj.com/articles/at-zara-fast-fashion-meets-smarter-inventory-1410884519 (accessed on 9 June 2015).
67. Special NRF Report: Decathlon Develops RFID to Revolutionise Stores and Back of House. The Retail Bulletin, 17 January 2014. Available online: http://www.theretailbulletin.com/news/special_

nrf_reportdecathlon_develops_rfid_to_revolutionise_stores_and_back_of_house_17-01-14/ (accessed on 9 June 2015).

68. Gilart-Iglesias, V.; Maciá-Pérez, F.; Marcos-Jorquera, D.; Mora-Gimeno, F.J. Industrial machines as a service: Modelling industrial machinery processes. In Proceedings of the 5th IEEE International Conference on Industrial Informatics, Vienna, Austria, 23–27 June 2007; pp. 23–27.
69. Andújar-Montoya, M.D.; Gilart-Iglesias, V.; Montoyo, A.; Marcos-Jorquera, D. A construction management framework for mass customisation in traditional construction. Sustainability 2015, 7, 5182–5210.
70. Eriksson, H.E.; Penker, M. Business Modelling with UML; John Wiley & Sons: Hoboken, NJ, USA, 2001.
71. García-Chamizo, J.M.; Mora-Pascual, J.; Mora-Mora, H.; Signes-Pont, M.T. Calculation methodology for flexible arithmetic processing. In Proceedings of the IFIP International Conference on Very Large Scale Integration (VLSI-SOC), Darmstadt, Germany, 1–3 December 2003; pp. 350–355.
72. Mora-Mora, H.; Mora-Pascual, J.; García-Chamizo, J.M.; Jimeno-Morenilla, A. Real-time arithmetic unit. Real Time Syst. 2006, 34, 53–79.
73. Mora-Mora, H.; Mora-Pascual, J.; Signes-Pont, M.T.; Romero, J.L.S. Mathematical model of stored logic based computation. Math. Comput. Model. 2010, 52, 1243–1250.
74. Paul, A.S.; Wan, E.A. Wi-Fi based indoor localization and tracking using sigma-point Kalman filtering methods. In Proceedings of the IEEE/ION Position, Location and Navigation Symposium, Monterey, CA, USA, 5–8 May 2008; pp. 646–659.
75. Dwoskin, E. What secrets your phone is sharing about you businesses use sensors to track customers, build shopper profiles. The Wall Street Journal, 13 January 2014. Available online: http://www.wsj.com/articles/SB10001424052702303453004579290632128929194 (accessed on 9 June 2015).
76. Nikitin, P.V.; Rao, K.V.S. Performance limitations of passive UHF RFID systems. In Proceedings of the IEEE Antennas and Propagation Society International Symposium, Albuquerque, NM, USA, 9–14 July 2006; pp. 1011–1014.

77. Li, N.; Becerik-Gerber, B. Performance-based evaluation of RFID-based indoor location sensing solutions for the built environment. Adv. Eng. Inform. 2011, 25, 535–546. [Google Scholar] [CrossRef]
78. Das, R.; Harrop, P. RFID Forecasts, Players and Opportunities 2014–2024; IDTechEx: Cambridge, MA, USA, July 2014. Available online: http://www.centrenational-rfid.com/docs/users/file/RFID_Forecasts_2014_2024.pdf (accessed on 9 June 2015).
79. Juels, A. RFID security and privacy: A research survey. IEEE J. Sel. Areas Commun. 2006, 24, 381–394.
80. Juels, A.; Rivest, R.L.; Szydlo, M. The blocker tag: Selective blocking of RFID tags for consumer privacy. In Proceedings of the ACM Conference on Computer and Communications Security, Washington, DC, USA, 27–30 October 2003; pp. 103–111.
81. RFIDControls. Available online: http://www.rfidcontrols.com/ (accessed on 9 June 2015).
82. Hohpe, G.; Woolf, B. Enterprise Integration Patterns, 1st ed.; Addison-Wesley: Boston, MA, USA, 2003.
83. Sivrikaya, F.; Yener, B. Time synchronization in sensor networks: A survey. IEEE Netw. 2004, 18, 45–50.
84. Zhang, F.; Deng, G.Y. Probabilistic time synchronization in wireless sensor networks. Int. Conf. Wirel. Commun. Netw. Mob. Comput. 2005, 2, 980–984.
85. Erl, T. SOA Design Patterns, 1st ed.; Prentice Hall: Upper Saddle River, NJ, USA, 2009.
86. Cubietruck. Available online: http://www.cubietruck.com/ (accessed on 9 June 2015).
87. Nikitin, P.V.; Rao, K.V.S. Antennas and propagation in UHF RFID systems. In Proceedings of the IEEE International Conference on RFID, Las Vegas, NV, USA, 16–17 April 2008; pp. 277–288.
88. Chen, Z.N.; Qing, X.; Chung, H.L. A universal UHF RFID reader antenna. IEEE Trans. Microw. Theory Tech. 2009, 57, 1275–1282.

CHAPTER 9

Creating TUIs Using RFID Sensors—A Case Study Based on the Literacy Process of Children with Down Syndrome

Janio Jadán-Guerrero [1,*], Luis Guerrero [1], Gustavo López [1], Doris Cáliz [2] and José Bravo [3]

[1] Research Center for Communication and Information Technologies, Universidad de Costa Rica CITIC-UCR, Ciudad Universitaria Rodrigo Facio, San José 2060, Costa Rica

[2] Universidad Politécnica de Madrid, Campus de Montegancedo, Madrid 28660, Spain

[3] MAmI Research Lab, Universidad de Castilla La Mancha, Ciudad Real 13071, Spain

ABSTRACT

Teaching children with intellectual disabilities is a big challenge for most parents and educators. Special education teachers use learning strategies to develop and enhance motivation for complex learning tasks. Literacy acquisition is an essential and life-long skill for a child with intellectual disabilities. In this context, technology can support specific strategies that will help children learn to read. This paper introduces a Tangible

User Interface (TUI) system based on Radio Frequency Identification (RFID) technology to support literacy for children with Down syndrome. Our proposed system focuses on the integration of RFID tags in 3D printed objects and low cost toys. The paper describes the experience of using some materials covering the tags and the different problems related to the material and distance of radio wave propagation. The results of a preliminary evaluation in a special education institution showed that the system helps to improve the interaction between teachers and children. The use of a TUI seems to give a physical sensory experience to develop literacy skills in children with Down syndrome.

KEYWORDS

tangible user interface; RFID; literacy; Down syndrome; interaction

1. INTRODUCTION

An appropriate education from an early stage in their life is critical to ensuring the integration of anyone into society. Literacy development is a key factor in this. Literacy starts at home and continues both in school and throughout life. In the case of children with Down syndrome, literacy development does not follow the traditional path (i.e., mild or moderate Down syndrome causes difficulties in the reading and writing learning processes) [1]. The lack of reading and writing abilities may cause difficulties in their future everyday activities, such as using public transportation or buying groceries [2].

Several studies report that children with Down syndrome can develop reading and writing abilities for concrete daily activities [1,2,3,4]. Moreover, it is possible to enhance speaking, reading, listening, and writing skills with the help of specific literacy methods [4,5].

In [4] the authors proposed a literacy method designed for children with Down syndrome. The method is based on a Global Method [2], which teaches children to read and recognize words as a whole, rather than breaking the word into individual letters or groups of letters. The method incorporates words and pictograms that are repetitively shown to the children.

Regardless of their condition, children are active learners who utilize all of their senses while exploring the world around them [5]. The incorporation of technology has transformed how children interact with and learn about their environment [6]. It has also impacted the way teachers create new teaching and learning strategies using computers, tablets, and other emerging technologies [7].

Designing technological interfaces to be used by children with disabilities is not an easy task. The main difficulty is the population heterogeneity both between different disabilities and within people who have the same disability [2]. Several academic researchers have focused their efforts on the design of technology made for children with disabilities [7,8,9,10] and the use of Tangible User Interfaces (TUIs) is one of the main trends in these technological tools.

Researchers agree that TUIs to improve children's cognitive and motor skills through computational devices. Moreover, the use of TUIs provides interaction with objects of various textures, shapes, sizes, and colors, which helps engage the child in the learning process [11,12,13].

This paper presents a TUI-based system which incorporates RFID technology. The TUIs were created using 3D printing and low cost toys. Our system is based on a literacy method for children with Down syndrome proposed by Troncoso and Del Cerro [4]. Moreover, we used a crowdsourcing approach to the creation of material and stored digital information in a Learning Objects Metadata (LOM) repository [14]. This information can be accessed by a Graphic User Interface (GUI) or printed in paperboard cards. This feature allows any teacher to both create materials and share them. The GUI and TUI prototypes were developed with the support of five special education teachers.

We performed two evaluations, testing the RFID technology with TUIs made of different materials (e.g., plastic, rubber, metal, and cardboard) in order to assess if the materials had any significant effect on RFID performance. The system was assessed in an experimental phase, conducted with five teachers and twelve children with Down syndrome.

The rest of this paper is organized as follows: Section 2 presents a background on literacy method for children with Down syndrome, TUI systems, RFID technology and related work of TUI systems for literacy. Section 3 describes the design and development strategies of our proposal. Section 4 examines the process to integrate RFID tags in objects and the evaluation with different materials. Section 5 describes the system evaluation in a real context. Section 6 provides a discussion of the results and gives some guidelines to enhance the use of our TUI system. Finally, Section 7 presents the conclusions and possible future extensions.

2. BACKGROUND AND RELATED WORK

There are many forms of intellectual disabilities; some common examples include: Down syndrome, autism, dementia, and Traumatic Brain Injury (TBI). Clinical diagnosis may also include less severe cognitive conditions such as dyslexia (an inability to read properly), dysgraphia (a difficulty with syntax), dyscalculia (an inability in math reasoning), and Attention Deficit Disorder (ADD, a difficulty with sustaining attention).

Some of the disabilities mentioned above are characterized by significant limitations both in intellectual functioning and in adaptive behavior. Intellectual functioning involves attention problems, memory deficit, slow learning rate, language-use difficulties and lack of motivation. Adaptive behavior involves self-care, daily living skills, social development and challenging behavior [15].

Teaching reading skills to children with intellectual disabilities is a challenge for teachers because learning abilities vary greatly depending on the intellectual disability. Additionally, differences between children who have the same disability can be vast [2,8,10,16].

Previous works suggest that early reading activities encourage the progress of children with intellectual disabilities [1,3,17]. Moreover, there is a trend in the development of methods and innovative interactive technologies for literacy acquisition of children with Down syndrome [4,5,6,7]. The following subsections describe a method for teaching reading to children with Down syndrome, introduce TUIs and RFID technologies and relate these concepts.

2.1. Literacy Method for Children with Down Syndrome

Children with Down syndrome typically learn to read the same way as all children. The difference is that they learn at a slower rate and their instruction must match this pace. They also need to develop comprehension and retention skills [1,2,3].

Troncoso and Del Cerro proposed a literacy method designed for children with Down syndrome [4] in which children recognize the meanings of written symbols by association, the same way they do with logos or brands. This method focuses on the development of many cognitive skills, including: attention and concentration; association of words and symbols with objects; perception and discrimination; identification of similarity and difference; classification of objects to see order or relation; and developing concepts, such as space, size, and shape [2].

The method begins with vocabulary that the children understand and are familiar with (e.g., animals, food, toys, and favorite places). Moreover, it utilizes pictograms and the written word on 15 × 10 cm cards. The font of the letters on the card is Script MS Bold, the color is red, and it is printed as large as possible. Figure 1 shows two examples of cards.

Reading activities begin by using cards in different ways, such as picture-to-picture (matching two pictures), word-to-picture (matching to make sure the child understands what he/she is reading), and word-to-word (matching the phonetics of the word). Teachers use all of these cards to develop memory, attention, association, discrimination, and denomination skills. While the method has complex resources, our research focuses only on the flashcards for reading.

Figure 1. Troncoso and Del Cerro's method cards.

.

2.2. Tangible User Interface

Ulmer and Ishii define TUIs as systems that use physical artefacts for representing and controlling digital information [11]. TUIs rely on a balance between physical and digital representations. Moreover, they help empower collaboration among children. TUIs facilitate learning and decision making through digital technology by taking advantage of our human ability to grasp and manipulate physical objects and materials [12].

TUIs take advantage of naturally acquired and developed physical object manipulation skills. Interacting with TUIs instead of traditional user interfaces also reduce the user's cognitive effort when interacting with a system [10].

These facts led us to believe that, since children learn a variety of skills by playing with physical objects, TUIs would be a "natural" form that could be used to interact with computers. Furthermore, TUIs have great potential to be applied to education [18,19]. Our proposal uses TUIs to help children with Down syndrome develop reading abilities by encouraging them to use their sensorial and cognitive capacities.

2.3. Radio Frequency Identification

Radio Frequency Identification (RFID) is an automatic identification technology that transmits the identity (ID) of an item wirelessly by using radio waves, tags and readers. When a tag passes through a field covered by a reader, it transmits a unique serial number. A tag can be passive or active. Both active and passive RFIDs use radio frequency energy to communicate between a tag and a reader; however, the method of powering the tags is different. Active RFID uses an internal power source (battery) within the tag to continuously power it. Passive RFID relies on radio frequency energy transferred from the reader to the tag to power the tag. [20,21]. Our system's implementation uses passive RFID tags.

Passive RFID tags are inexpensive (approximately $0.20 USD each) and their storage capacity is low (less than 256 bytes). The major advantage of passive RFID tags is their size, as they can be as small as 0.705 inches—approximately the size of an American dime. However, the use range of passive tags is small; depending on the tag size, it can range from a couple of inches to seven or eight inches. The main drawback of passive RFID tags is that any metal surface presents an impassable barrier [22].

The operating frequency of an RFID system directly influences its read range. Passive RFID requires stronger signals from the reader, and the signal strength returned from the tag is constrained to very low levels. RFID operation frequencies range from 125 KHz to 2.4 GHz [23]. In [24], the authors describe how interacting with different devices is beneficial to people with cognitive disabilities and present a system that use RFID technology to interact with games.

2.4. TUIs for Early Literacy

Several TUIs that support the educational process have been presented in literature. This section describes some of the most relevant projects.

RoyoBlocks is a system developed to allow children to build sentences. RoyoBlocks consists of 60 wooden blocks and a stuffed monkey. Each block is a word and has an embedded RFID tag. The monkey contains an Arduino powered RFID reader, which identifies the tag of each block and plays audio of the word pronunciation through a speaker. This system is offered to provide young children an opportunity to develop early literacy skills. Only preliminary testing has been performed with RoyoBlocks [9].

LIT KIT is a portable system that supports children book reading. It uses Arduino technology which communicates with Sifteo Cubes [25] to control the actions of the multi-media, architectural robot. Sifteo Cubes are blocks (based on the timeless play patterns of Legos, building blocks, and domino tiles), which communicate wirelessly and respond to each other. LIT KIT experience provides children an opportunity to interact with digital and robotic artifacts within a real, physical environment that transforms words into worlds [26].

READING GLOVE is a wearable RFID reader and a set of common objects with the tags under or inside objects. Children select an object and pick it up in order to listen the word's pronunciation. The system engages readers in tangible and audio-based interaction [27].

TABLE TOP is a system that uses an inverted projection of a video beam. The system also uses tangible objects that are recognized by external tags and a camera. The tangible interfaces are physical objects that represent the images or words shown [7,10]. A research study describes how tabletops are an interesting approach to computer supported collaborative learning and the authors listed issues that must be addressed when using this type of interface [28].

E-DU BOX is a computer application and a tangible interface that uses a special pen-shaped mouse that vibrates according to some situations defined by the educator in the reading process. Feedback is also provided by a tangible, interactive and animated e-du agent, these agents are able to move and speak to students. All of these components use Bluetooth-based connection [29].

CO-STICAP (Stimulating Collaborative Cognitive Capabilities) is a multi-device based on distributed user interfaces and games systems. It aims to provide cognitively stimulating activities for children with Attention Deficit Hyperactivity Disorder (ADHD). It uses a laptop, a smartphone and a tangible object coded with NFC (Near-Field Communication) labels [30].

TICLE (Tangible Interface for Collaborative Learning Environments) is a platform that uses computer vision techniques to track objects and map their movements into a graphical user interface. Children interact with concrete mathematical games, such as Hanoi Tower and Tangram, while the system maps their actions to the computer screen [31].

I/O Brush is a drawing tool, in the shape of a common paintbrush, with an embedded camera and touch sensors. This device allows users to capture colors and textures of surfaces and reproduce them on the drawing canvas (consisting of a large touchscreen and a back projection screen) [32].

All the systems mentioned above offer an environment to practice the educational concepts of learning-by-doing, which can enhance the efficiency of a child's learning process [17]. This information has allowed us to propose a system reflecting needs in our local context.

3. TANGIBLE USER INTERFACE DESIGN AND DEVELOPMENT STRATEGY

This section describes the design and development strategy we followed to build our proposed TUI system. The development processes was supported by five special education teachers. We used an Iterative User-Interface Design to develop prototypes and to consider some usability metrics, such as, easy to learn, efficient to use, easy to remember and pleasant to use [33]. We worked with teachers in exploring the use of the resources of the literacy method and how to adapt the technology to their needs.

The first prototype was based on the cards that the literacy method applies. We gathered the opinion of teachers and selected some concepts of study divided into four categories (fruit, animals, home and entertainment). We designed and printed a set of six cards for each category.

For the second prototype we designed a graphical user interface (GUI) running on web browser which could be accessed by computers o mobile devices. The user interface was designed to be intuitive and easy to configure for non-technical users. The information was organized in a database and used to create learning objects based on the standard IEEE Learning Objects Metadata (LOM) [14].

The third prototype based on TUI was the extension of the above prototypes. We analyzed the possibility to build the objects with a printer 3D and low cost toys. Next section details the process to integrate RFID technology to the objects. The backend of the system was developed using PHP and MySQL technology, and the frontend was developed using PHP, HTML5, CCS3 and JavaScript. This open-source technology allows the information to be shown in any browser of any device that is connected to the Internet. A Web service was developed to communicate with the software that identifies the tags. On the client side, the process unit was developed using Microsoft .Net technology. Figure 2 shows an architectural diagram of our system.

Three main modules compose the system:

The objects recognition tool. The tool has two RFID readers attached to the processing unit. It recognizes the RFID tag inside the object placed over it and sends a signal to display the associated word and pronunciation.

The recognizable objects set. We have several sets of 24 objects. These objects were redesigned in order to contain an RFID tag that recognizable by the system.

The graphic user interface. Tablets, smart phones, PC or even an interactive board with access to the Internet can be used to display the pictograms, words and pronunciations representing the physical objects.

The processing unit receives the ID for one of the RFID sensors. The software matches the received ID with the records in the repository through a Web service. The system not only displays a picture or pictogram representing the object but also presents the word and pronunciation associated with the object in each device with an open session.

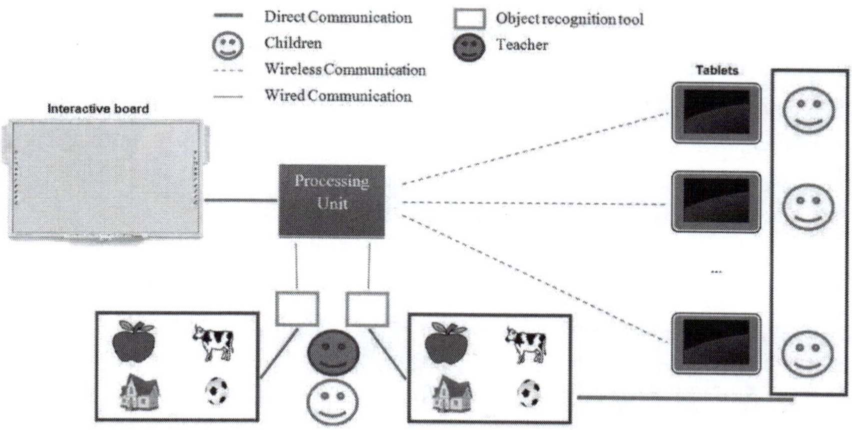

Figure 2. Diagram of the Tangible User Interface based on RFID sensors.

The processing unit has an administration module to configure the number of sessions and the name or number that identifies the teacher and children. Tablets or mobile devices can be connected via Web through the session number. The system registers information about the activities carried out by teachers and children, such as: the start and end time of each session, the number of objects selected by the teacher and the child, and the time between each identification step.

Our design of TUIs and interaction allows pairing between either one TUI set and one GUI or, one TUI set and multiple GUIs. This allows for simultaneous interaction. Moreover, as we used Mobile-Web technology in the development process, computers, tablets or interactive boards could be used to display the digital representation of recognized objects. This last feature was designed to be used in therapy sessions (teacher-child) or in a classroom.

In order to evaluate the design of the prototypes we carried out we conducted a thinking-aloud test [34] followed by an interview about the user experience with five special education teachers. The goal was to understand their experience while teaching children with Down syndrome. This includes the method, type of educational resources and teaching strategies they used to teach children how to read. For this purpose, we explained how each prototype works and asked them to verbalize their thoughts, simulating a real scenario. The test helped us to identify some problems related to the size of pictograms, the type and color of the words, the type of voice and the time for automatic pronunciation. The feedback of the test enabled us to improve the system with the following features:

(a) The color of the physical object are the same of the pictogram in the GUI.

(b) The color of the words in cards and GUI are red.

(c) The size of the pictogram and the words are big as possible.

(d) Each word is preceded by the article.

(e) The GUI has an option to change the font, color and size of the words.

(f) The voice of the system is female.

Figure 3 shows the final version of the system.

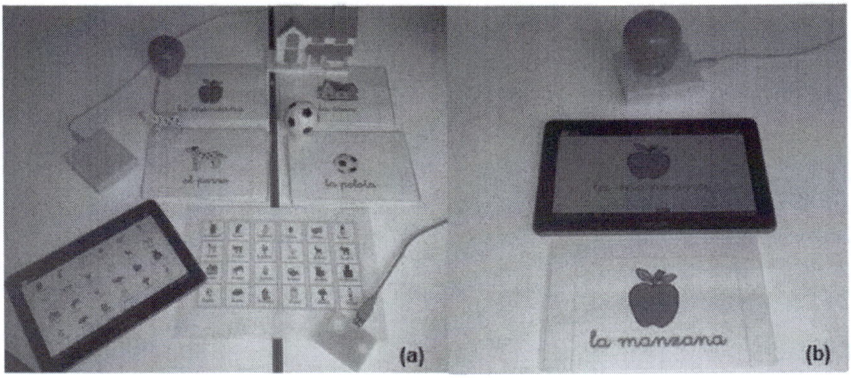

Figure 3. (a) Tangible User Interfaces for literacy; (b) Three interfaces (TUI, GUI and Card).

Our system differs from the solutions presented in the related work section because we used Wi-Fi technology. This feature enables long distance communications and portability. It also differs because we avoided using external tags or signs that modify the real appearance of the objects. We accomplished this by applying a digitalization and rapid manufacturing process (i.e., 3D scanning and printing to redesign the objects and embed the required tags to the toys) [21].

In the interviews about user experience, all teachers agreed that the system was intuitive and easy to use. Teachers pointed out that children's intellectual development takes place not only with the support of more capable peers or adults through social interaction, but also with the evolvement of social and cultural artifacts. This assertion is based on pedagogical theories studied by Vygotsky [35]. In this sense the interaction with technology could help to develop strategies for connecting abstract and concrete representations of concepts.

4. INTEGRATING RFID TAGS INTO OBJECTS

This section describes the process to integrate RFID tags into the tangible objects. We selected two RFID readers. The first was Phidget RFID reader [36] which has a set of three tags: (a) ABS Key Fob Blue; (b) Clothing Button-2.5 cm diameter; (c) Credit Card 4 × 3 cm. The second was Tertium icekey Hf reader [37] with a set of three tags: (d) HF tag Sticker 1.8 × 3.6 cm; (e) HF tag Sticker 4.3 × 4.3 cm; (f) Credit card 4 × 3 cm (see Figure 4). Each tag relies on a small chip, which is implanted in it and has a unique code of identification (ID). The tag can be attached or embedded to any kind of things. Then, an electronic scanner (RFID reader) uses radio signals to read or track the ID tag [20].

According to the manufacturers, passive tags require a strong RF field to operate and their effective range is limited to an area in close proximity to the RFID reader. The distance over which the RFID tag is usable is affected by things, such as: the tag shape and size, the materials being used in the area near the reader, and the orientation of the reader and tag with respect to each other. The smaller a tag is, the closer it must be to the reader to be detected [38]. As a result of declaring that the recognizable object could not have external tags as a key characteristic of our system, we considered two main aspects: the material of the object and the orientation of the tags.

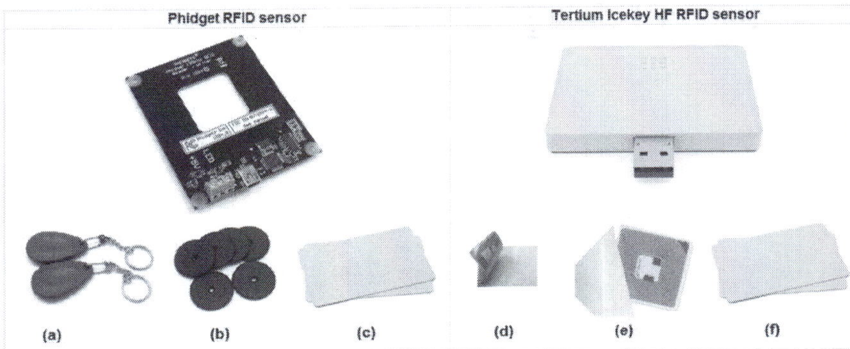

Figure 4. Passive RFID tags of Phidget RFID and Tertium icekey Hf readers. (a) ABS Key Fob Blue; (b) Clothing Button: 2.5 cm diameter; (c) Credit Card: 4 × 3 cm; (d) HF tag Sticker: 1.8 × 3.6 cm; (e) HF tag Sticker: 4.3 × 4.3 cm; (f) Credit card: 4 × 3 cm.

We created 3D printed objects with internal space for tags, and used low cost toys that could be disassembled. Figure 5a shows some toys with the embedded tags. Figure 5b shows the measuring process with the objects.

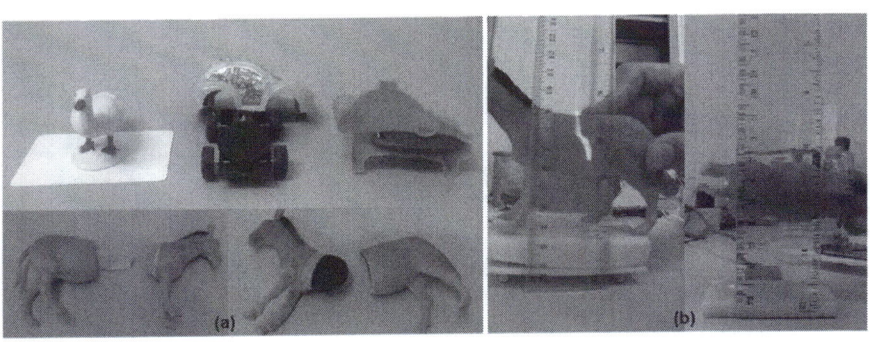

Figure 5. (a) Tags in toys and 3D printed objects; (b) Measure process with Tertium, icekey, Hf and Phidget RFID readers.

In order to measure how the effectiveness of the readers was affected by materials, we carried out a test with the following materials: aluminum, glass, iron, paperboard, plastic, porcelain, rubber and wood. The results of the measurements are shown in Figure 6 and Figure 7.

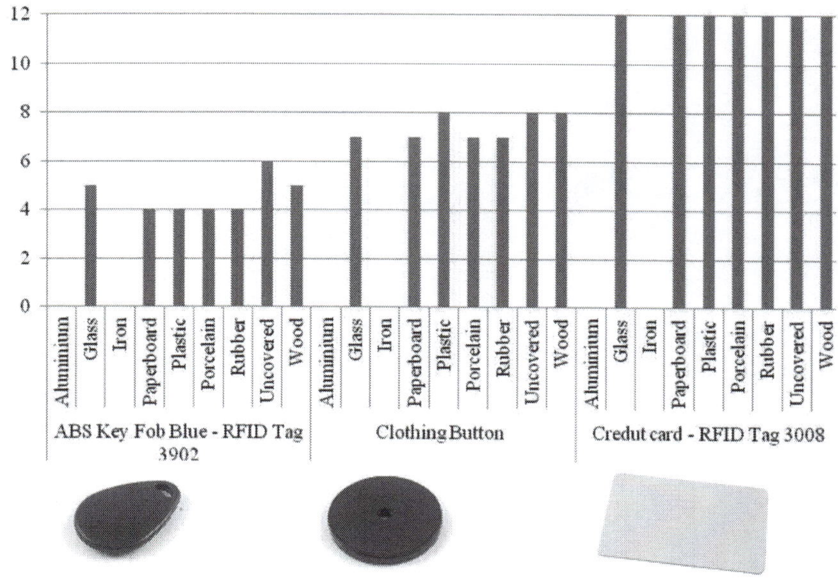

Figure 6. Distances registered using Phidget RFID sensor.

Based on the functionality of passive RFID tags [39,40] the measures were performed in such a way that the reader and the tag were parallel to each other. The three tags of each reader were placed inside or over each material and we performed two measurements for each one

One revised study mentioned that RFID transmits signals through electromagnetic waves; therefore, it is extremely sensitive to metals, liquids or shiny covers [41]. Our results confirmed this fact—Figure 6 and Figure 7 show that metallic surfaces were not recognized by any of our readers. We also confirmed that the distance of recognition is directly proportional of the size of the tags [42]. The base line was established testing uncovered tags, and they registered the maximum distance of 12 cm for Phidget and 10 cm for Tertium.

The results were similar for both readers. We noted that the presence of some materials, for example rubber, near tags can degrade the performance of the RFID, preventing them from being read. However, with glass, paperboard, plastic, porcelain and wood the measurements were constant. We observed that the position also influences in the recognition. The radio wave sensor works well when the tags are placed parallel, but the identification decreases when the tag is in a vertical position.

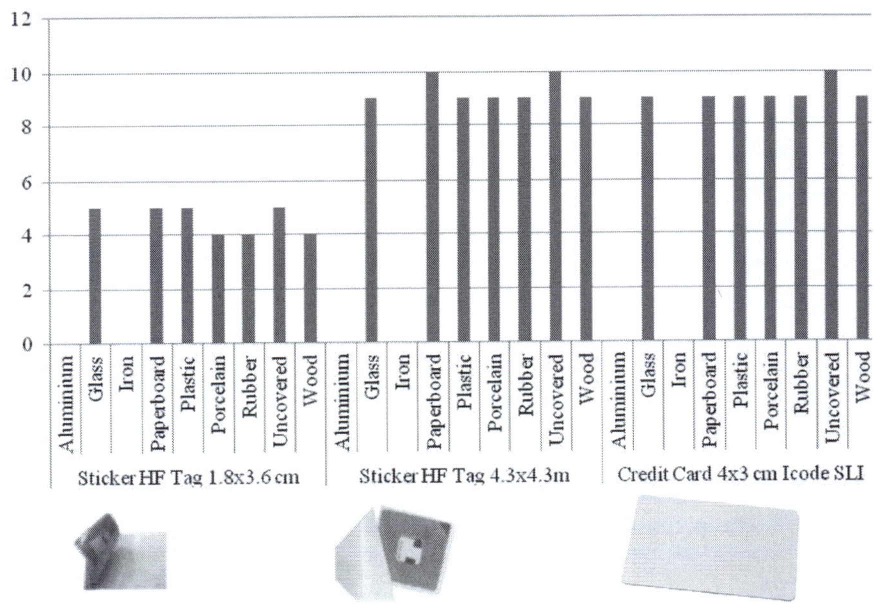

Figure 7. Distances registered using Tertium Icekey HF RFID sensor.

5. SYSTEM EVALUATION

Structured experiments were performed to assess the perceived usefulness and usability of our system in real scenarios. We carried out the evaluation in a special education institution with five teachers and twelve children with Down syndrome. To perform the evaluation, teachers were trained on how to use our system and they were who incorporated it into the learning process that they normally follow. Each child participated in two processes: (1) teacher-child, one teacher interacts with one child; (2) Self-learning, one child interacts individually with the system.

In the first scenario, the system was set up with the following components: one Laptop (Processing unit), two RFID readers, a set of 24 recognizable objects, and a tablet or interactive board. Each teacher carried out reading activities related to the method with each assigned child. The procedure was performed with three versions of prototypes: Card (paperboard cards), GUI (digital cards displayed on a tablet or interactive board) and TUI (Tangible RFID objects). The teacher selects one card and shows it to the child. The teacher read the associated word of the pictogram. The child repeats the pronunciation and identify the concept. In the GUI version the teachers works with a tablet or interactive board. The system shows a menu with the set of 24 pictograms. The teacher selects randomly one pictogram and the system reads the word. Next is the turn of the child. In the TUI version the teacher and the child interact with the set of 24 physical objects. The teacher selects randomly one tangible object and the system pronounces the name of the object when it is close of the sensor. The child repeats the task performed by the teacher.

In the second scenario, each child interacts with the three prototypes by himself. The prototype is selected randomly and each child is free to use the cards, the tablet and the tangible objects during 5 min each one. This activity was designed to observe a self-learning behavior. Figure 8 shows two examples of sessions: in Figure 8a, a child interacts with the tangible objects and a tablet; and in Figure 8b, another child interacts with the tangible objects and the interactive board.

Based on Hawthorne studies we made a shift from controlled experiments toward a complex social situation [43], i.e., we observed the behavior of both teachers and children with the system. The interaction with tangible objects was very natural for children. They experienced positive emotions and spent an enjoyable time with GUI and TUI.

Figure 8. (a) A child interacts with the tangible objects and a tablet; (b) A child interacts with the tangible objects and the interactive board.

Since the evaluation was not designed to obstruct the traditional learning process, we did not assess the children appreciation of the tool during the evaluation; however, their productivity increased, as a sign of motivation. Moreover, according to the teachers, they provided more and better feedback when using our system than in traditional lessons.

During the evaluation, two researchers observed the progress and noticed that:

Children required visual or auditory confirmation of the system when an object was recognized. We implemented a beep to confirm recognition.

Children had difficulties with some objects, specifically 4-legged animals (e.g., cow, dog, and donkey). When the children placed them in a vertical position, the RFID tag was not close enough to be recognized. We used two readers to avoid this problem. Figure 9 shows the configuration of two RFID readers.

We observed the engagement and interaction of children with the tangible objects. This engagement may result from motivation of toys or objects which are familiar for children with Down syndrome [13]. The teachers also expressed motivation when using the system and its enhanced capabilities to create new learning and reading materials. The toys and other objects were attractive to create stories and strengthen language skills.

6. DISCUSSION

After the validation process, all of the teachers agreed that the system is useful in enhancing the first stages of the literacy process. One of the teachers mentioned that the system could not only help children with intellectual disabilities, but also children with visual disabilities when touching and recognizing the objects. The teacher believed that the adaptation of novel technology is necessary for the integration of children with disabilities in the classroom.

Teachers also thought that tangible interfaces emphasize the connection between the body and cognition, facilitating thinking through physical actions. Tangible media is a complement to literacy method because it gives digital information and physical shape to the studied words.

It is important to mention that most of the participants in the evaluation process pointed the fact that the system could be used to help the literacy process for any child, including children without disabilities.

A key component of learning strategies to develop reading skills is interaction [13,16]. In this sense, we analyzed two learning strategies: teacher-child and self-learning. In each strategy we carried out a random tests during each session. Table 1 shows the results of the level of interaction with the system between teachers and children.

Table 1. Results of the level of interaction

Learning Strategy	Interface	Children	Teachers
Teacher-child	Card	low	high
	GUI	medium	medium
	TUI	high	Low
Self-learning	Card	low	NA
	GUI	medium	NA
	TUI	high	NA

The level of interaction children had with TUI was high in relation to that of GUI and Card. It was evident that children used not only verbal language but also the sensorial-perceptive stimulation. In the GUI, the interaction was shared between the child and the teacher. However the interaction with the traditional interface (Cards) was low by children.

In the evaluation context we observed some problems with the identification process. Some tangible objects were not easily identified because of the position of the tag. However, children learned to find the best position by moving the object until one of the readers beeped. We also observed that teachers who participated in the study had difficulties in positioning some objects. This led us to think that we could improve the identification of the objects. Based on previous works on using RFID technology in the classroom [44,45,46], we propose a stage with the two readers placed in horizontal and vertical positions, as depicted in Figure 9.

The simulation of a real stage could help the children identify the tangible objects easier. Teachers could create learning strategies to develop skills, such as: attention and concentration; associating words and symbols with objects; sight-word recognition; and phonics. Teachers could also help introduce and develop new vocabulary; perception and discrimination; identifying similarity and differences; classifying objects; seeing order or relationships; developing concepts like space, size, and shape; and exploring, manipulating, and using creative imagination [34,47].

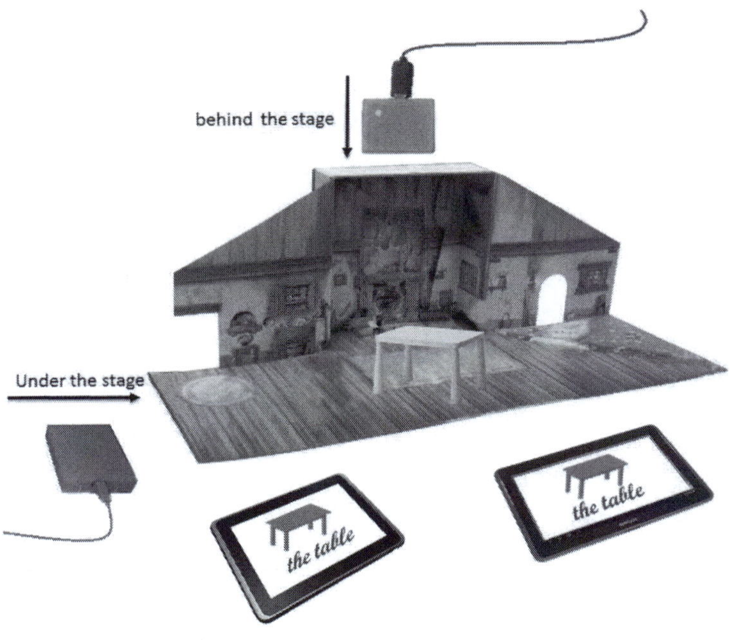

Figure 9. Using two sensors to facilitate object recognition.

Children with Down syndrome will usually be delayed in their language comprehension compared to other children of the same age. It is important for them to start with vocabulary they can understand and that words or objects be familiar to them. Both teachers and parents need to spend time with the children to enhance their reading skills and the system could contribute to that. Reading practice will also help develop working memory skills. The ability to read and write facilitates understanding of general knowledge from the school curriculum and supports the necessary skills for problem solving in real life.

7. CONCLUSIONS

In this paper, the use of RFID technology was analyzed when implementing a set of tangible objects to support literacy. Different problems related to the material and distance of radio wave propagation were evaluated.

We have shown that Tangible User Interfaces can be created from common objects in our daily environment and that this integration can be important to design new educational technologies that support literacy.

The results of the prototype evaluation showed that the system, not only assists the teacher, but also helps improve the interaction between children and non-traditional user interfaces. It is important to mention that the educational inclusion of children with intellectual disabilities works better when all of their classmates are involved in same activities. A technological system for supporting this process can be used but can never replace the teacher guide.

We found that a system with a Tangible User Interface, like the one we developed, can support teachers or parents working with children with intellectual disabilities, especially if the system supports a well-known method, like the Troncoso and Del Cerro literacy method. We want to highlight that the final users of our system are the teachers; our system supports those who are experts in teaching literacy to children with learning disabilities.

Depending on the cognitive deficit of the student, more features can be easily implemented and incorporated into the system. For instance, the card on the tablet could include a sound or video. In the same way, using their finger, children can paint over or follow the written word on the tablet to introduce writing. These new features are scheduled for a future version of the prototype and more evaluations and experiments will be conducted.

In educative systems the focus is not the technology, but the interaction it facilitates and its ultimate learning effects. This interaction can also play an important role in the learning of new skills because children learn by experiencing the environment through their senses. The use of multisensory strategies can help to scaffold literacy learning [16].

Kubicki, Lepreux and Kolski [48,49], presented TangiSense, a table that allows interactions with tangible objects. We consider that by incorporating TangiSense capabilities and our system, teacher could engage children in more complex and complete activities, and allow better monitoring of the situation at all times.

ACKNOWLEDGMENTS

We would like to thank the Down institution ASINDOWN. We thank our research colleagues in DISC (Departamento de Sistemas Informáticos y Computación, Universidad Politécnica de Valencia). This research was supported by SENESCYT (Secretaría Nacional de Educación Superior, Ciencia y Tecnología, Ecuador), UTI (Universidad Tecnológica Indoamérica, Ecuador), PCI-UCR (Posgrado en Computación e Informática de la Universidad de Costa Rica), CITIC-UCR (Centro de Investigación en Tecnologías de la Información y la Comunicación, Universidad de Costa Rica) and SEP (Sistema de Estudios de Posgrado de la Universidad de Costa Rica).

AUTHOR CONTRIBUTIONS

Janio Jadán developed the system, carried out the experiments and wrote the initial version of the manuscript. Luis Guerrero was the director of this research who supported with Human-Computer Interaction aspects. Gustavo López was the designer of the first version of the prototype and also contributed to the structure of the manuscript. Doris Cáliz contributed to the experiments and also performed the literature review. José Bravo evaluated the results and contributed to the contents and structure of the manuscript.

REFERENCES

1. Abbeduto, L.; Warren, S.F.; Conners, F.A. Language development in Down syndrome: From the pre-linguistic period to the acquisition of literacy. Ment. Retard. Dev. Disabil. Res. Rev. 2007, 13, 247–261.
2. Buckley, S.J.; Bird, G. Teaching children with Down syndrome to read. Down Syndr. Res. Pract. 1993, 1, 34–39.
3. Buckley, S.J.; Beadman, J.; Bird, G. Reading and writing for children with Down syndrome (5–11 years). Down Syndr. Issues Inf. 2001.
4. Troncoso, M.V.; del Cerro, M.M. Síndrome de Down: Lectura y Escritura; Santander Spain: Masson, Spain, 1999.
5. Badillo, V. Using picture communication symbols (PCS) for teaching to read and write children with Down's syndrome. Int. J Educ. Sci. Res. 2014, 4, 1–8.
6. Marco, J.; Cerezo, E.; Baldassarri, S. Bringing tabletop technology to all: Evaluating a tangible farm game with kindergarten and special needs children. Pers. Ubiquit. Comput. 2013, 17, 1577–1591.
7. Haro, B.; Santana, P.; Magaña, M. Developing reading skills in children with Down syndrome through tangible interfaces. In Proceedings of the 4th Mexican Conference on Human-Computer Interaction, Mexico City, Mexico, 3–5 October 2012; pp. 28–34.
8. Hsin, C.; Li, M.; Tsai, C. The influence of young children's use of Technology on their earning: A review. Educ. Technol. Soc. 2014, 17, 85–99.

9. Kleiman, J.; Pope, M. RoyoBlocks: An exploration in tangible literacy learning. In Proceedings of the 12th International Conference on Interaction Design and Children, New York, NY, USA, 24–24 June 2013; pp. 543–546.
10. Sitdhisanguan, K.; Chotikakamthorn, N.; Dechaboon, A.; Out, P. Using tangible user interfaces in computer-based training systems for low-functioning autistic children. Pers. Ubiquit. Comput. 2012, 16, 143–155.
11. Ullmer, B.; Hiroshi, I. Emerging frameworks for tangible user interfaces. IBM Syst. J. 2000, 39, 915–931.
12. Ishii, H. The tangible user interface and its evolution. Commun. ACM 2008, 51, 32–36.
13. Xie, L.; Antle, A.N.; Motamedi, N. Are tangibles more fun? Comparing children's enjoyment and engagement using physical, graphical and tangible user interfaces. In Proceedings of the 2nd International Conference on Tangible and Embedded Interaction, Bonn, Germany, 18–20 February 2008; pp. 191–198.
14. Ochoa, X.; Klerkx, J.; Vandeputte, B.; Duval, E. On the use of learning object metadata: The GLOBE experience. Lect. Notes Comput. Sci. 2011, 6964, 271–284.
15. King, B.H. Intellectual disability: Understanding its development, causes, classification, evaluation, and treatment. J. Am. Med. Assoc. 2008, 299, 1194.
16. Neumann, M.; Hyde, M.; Neumann, D.; Hood, M.; Ford, R. Multisensory Methods for Early Literacy Learning; Nova Science Publishers: Hauppauge, NY, USA, 2012; pp. 197–216.
17. Epstein, A.; Hohmann, C.; Hohmann, M. How Young Children Learn to Read in High/Scope Programs. Available online: http://www.highscope.org/file/NewsandInformation/PositionPapers/Literacy2.pdf (accessed on 5 May 2015).
18. Xu, D.; Mazzone, E.; MacFarlane, S. Informant design with children—designing children's tangible technology. In Proceedings of the 6th International Conference on Interaction Design and Children, Aalborg, Denmark, 6–8 June 2007; pp. 209–212.
19. O'Malley, C.; Fraser, D.S. Literature review in Learning with tangible technologies. NESTA Futurelab Rep. 2005, 12, 12–17.

20. Expósito, I.; Cuiñas, I. Exploring the limitations on RFID technology in traceability systems at beverage factories. Int. J. Antennas Propag. 2013, 2013, 1–6.
21. Angerer, C.; Langwieser, R.; Rupp, M. Evaluation and exploration of RFID systems by rapid prototyping. Pers. Ubiquit. Comput. 2012, 16, 309–321.
22. Dolgui, A.; Proth, J.M. Supply Chain Engineering: Useful Methods and Techniques; Springer-Verlag: London, UK, 2010; pp. 163–194.
23. Nekoogar, F.; Dowla, F. Ultra-Wideband Radio Frequency Identification Systems; Springer-Verlag: Berlin, Germany, 2012; pp. 1–157.
24. De La Guía, E.; Lozano, M.D.; Penichet, V.M.R. Improving the capabilities of people with cognitive disabilities with an interactive system based on distributed user interfaces. In Proceedings of the 2nd Workshop on Distributed User Interfaces: Collaboration and Usability. In Conjunction with CHI 2012 Conference, Austin, TX, USA, 5 May 2012.
25. Sifteo Inc. Sifteo Cubes. Available online: https://www.sifteo.com/cubes (accessed on 19 June 2015).
26. Schafer, G.; Green, E.; Walker, I.; Lewis, E.; Fullerton, S.; Soleimani, A.; Norris, M.; Fumagali, K.; Zhao, J.; Allport, R.; et al. Designing the LIT KIT, an interactive, environmental, cyber-physical artifact enhancing children's picture-book reading. In Proceedings of the 12th International Conference on Interaction Design and Children, New York, NY, USA, 24–27 June 2013; pp. 281–284.
27. Tanenbaum, K.; Tanenbaum, J.; Antle, A.; Bizzocchi, J.; Seif el-Nasr, M.; Hatala, M. Experiencing the reading glove. In Proceedings of the Fifth International Conference on Tangible, Embedded, and Embodied Interaction, Funchal, Portugal, 22–26 January 2011; pp. 137–144.
28. Dillinbourg, P.; Evans, M. Interactive tabletops in education. Int. J. Comput. Support. Collab. Learn. 2011, 6, 491–514.
29. Brotto, A.; Pontual, T.; Gomes, A.; Eduardo, C.; Rodrigues, M.; Sonnino, R. E-du box: Educational multimedia with tangible-enhanced interaction. In Proceedings of the 7th ACM Conference on Designing Interactive Systems, Cape Town, South Africa, 25–27 February 2008; pp. 139–146.

30. De la Guía, E.; Lozano, M.; Penichet, V. Co-StiCap: Sistema basado en interfaces de usuario distribuidas y tangibles para mejorar las capacidades cognitivas en niños con TDAH. In Proceedings of the XIV Congreso Internacional de Interacción Persona-Ordenador, Madrid, Spain, 17–20 September 2013; pp. 62–68. (In Spanish).
31. Scarlatos, L.L.; Dushkina, Y.; Landy, S. TICLE: A tangible interface for collaborative learning environments. In Proceedings of the Extended Abstracts of the SIGCHI Conference on Human Factors in Computing Systems–CHI'99, Pennsylvania, PA, USA, 15–20 May 1999; pp. 260–261.
32. Ryokai, K.; Marti, S.; Ishii, H. I/O Brush: Drawing with everyday objects as ink. In Proceedings of the SIGCHI Conference on Human Factors in Computing Systems, Vienna, Austria, 24–29 April 2004; pp. 1–8.
33. Nielsen, J. Iterative user interface design. IEEE Comput. 1993, 26, 32–41. [Google Scholar] [CrossRef]
34. Plaisant, C.; Shneiderman, B. Designing the User Interface: Strategies for Effective Human Computer Interaction, 5th ed.; Addison-Wesley Publisher: Boston, UK, 2010; pp. 1–650. [Google Scholar]
35. Vygotski, L.; Cole, M. Mind in Society: The Development of Higher Psychological Processes; Harvard University Press: Cambridge, MA, USA, 1978; pp. 1–101.
36. Phidgets Inc. 1024_0—PhidgetRFID Read-Write. Available online: http://www.phidgets.com/products.php?product_id=1024 (accessed on 18 June 2015).
37. Tertium Technology—ICEKEY HF. Available online: http://www.tertiumtechnology.com/icekey_hf.php (accessed on 18 June 2015).
38. Heidt, M.; Pfeiffer, L.; Bischof, A.; Rosenthal, P. Tangible Disparity-Different Notions of the Material as Catalyst of Interdisciplinary Communication. In Human-Computer Interaction. Theories, Methods, and Tools; Springer International Publishing: Crete, Greece, 2014.
39. Koski, E.; Bjorninen, T.; Ukkonen, L.; Sydanheimo, L. Radiation Efficiency Measurement Method for Passive UHF RFID Dipole Tag Antennas. IEEE Trans. Antennas Propag. 2013, 61, 4026–4035.

40. Tentori, M.; Cornejo, R.; Favela, J. Sentient displays to connect elders with cognitive disabilities to the digital ERA of social interaction. In Proceedings of the First International Workshop on Interactive Systems in Healthcare, Atlanta, GA, USA, 11 April 2010; pp. 161–164.
41. Shu-hsien, T.; Chien-ju, Ch. A Study on the Influence of RFID Tagging on Circulation Services nd Collection Management: A Case Study of the Taipei Public Library. Available online: http://cdn.intechopen.com/pdfs-wm/18100.pdf (accessed on 3 May 2015).
42. Kriara, L.; Alsup, M.; Corbellini, G.; Trotter, M.; Griffin, J.; Mangold, S. RFID shakables: Pairing radio-frequency identification tags with the help of gesture recognition. In Proceedings of the ACM Conference on Emerging Networking Experiments and Technologies, Santa Barbara, CA, USA, 9–12 December 2013; pp. 327–332.
43. Obrenović, Ž. The Hawthorne studies and their relevance to HCI research. Interact. Mag. 2014, 21, 46–51.
44. Jones, A.; Dontharaju, S.; Tung, S.; Mats, L.; Hawrylak, P.; Hoare, R.; Cain, J.; Mickle, M. Radio frequency identification prototyping. ACM Trans. Des. Autom. Electron. Syst. 2008, 13, 1–22.
45. Bravo, J.; Hervás, R.; Chavira, G. Ubiquitous computing in the classroom: An approach through identification process. J. Univers. Comput. Sci. 2005, 11, 1494–1504.
46. Bravo, J.; Hervás, R.; Nava, S.; Chavira, G.; Parras, J.; Delgado, M.; Vazquez, A.; Terán, F.; Sanchez, I.; Viñuela, U.; et al. Towards the everyday computing in the classroom through RFID. In Computers and Education; Springer Netherland: Dordrecht, Netherlands, 2007; pp. 143–153.
47. Figueiredo, L.; Dos Anjos, R.; Lindoso, J.; Edvar, N.; Rafael, R.; Silva, M.; Teichrieb, V. Bare hand natural interaction with augmented objects. In Proceedings of the 2013 IEEE International Symposium on Mixed and Augmented Reality (ISMAR), Adelaide, SA, USA, 1–4 October 2013; pp. 1–6.
48. Kubicki, S.; Lepreux, S.; Kolski, C. Evaluation of an interactive table with tangible objects: Application with children in a classroom. In Proceedings of the 2nd Workshop on Child Computer Interaction UI Technologies and Educational Pedagogy, Vancouver, BC, Canada, 7–12 May 2011; pp. 1–4.

49. Kubicki, S.; Lepreux, S.; Kolski, C. RFID-driven situation awareness on TangiSense, a table interacting with tangible objects. Pers. Ubiquit. Comput. 2012, 16, 1079–1094.

CHAPTER 10

A systematic review of RFID applications and diffusion: key areas and public policy issues

Kwangho Jung[1*] and Sabinne Lee[2]

1 Korea Institute of Public Affairs, Graduate School of Public Administration of Seoul National University

2 Graduate School of Public Administration of Seoul National University

ABSTRACT

RFID applicants called as e-ID, smart tag, and contactless smart card are being applied to numerous areas in our daily life, including tracking manufactured goods, currency, and patients to payments systems. To review these various applications of RFID is important to exploring not only ongoing e-governance issues such as digital identification, delivery process, and governance but also business oriented application areas like supply chain. Through a systematic review methodology from 111 previous studies about RFID technology for public sector, we found six key areas of RFID applications: defense and security, identification, environmental applications, transportation, healthcare and welfare, and agriculture-livestock. We also suggest that the diffusion and applications of RFID can involve unexpected disadvantages including technological deficiency, uncertain benefits, dubious transparency, uncomfortable privacy issue, and unequal distribution of digital power and literacy. Further research on RFID impact includes not only various theoretical issues of but also legal and managerial problems. Rigorous research is required to explore what factors are critical to adopt and implement new RFID applications in terms of technology governance and digital literacy. Massive data driven research is also expected to identify RFID performance in government agencies and various industry sectors.

KEYWORDS

RFID Digital identification Digital delivery Smart tag Contactless smart card

1. BACKGROUND

RFID technology has been widely implemented all over the world and its impact on our daily life is very diverse and massive (Li et al., 2006; Wyld, 2005). Those diverse areas of RFID application include logistical tracking, monitoring and maintenance of products, product safety and information, and payment process. Today many governments around the world in both developed[1] and developing[2] countries are trying to apply it for various areas from tracking manufactured goods, currency, and patients to securing sagety of payments systems. Massive RFID applications around all the industry sectors and countries are expected to generate a huge potential benefits for sustainable efficient energy infrastructure, transportation safety, and health care. Over the past 50 years, RFID technology went through innovations and progressions to become a more efficient and effective gadget for human beings as well as effective solutions of technical and organizational problems in various industry sectors. However, key issues of appropriate ICT technology, governing networks among RFID domains, standardization requirement, and privacy still remain unsolved[3].

We review previous literature about RFID technology used in public sectors in order to identify what has been done and found to suggest policy implications and further research agenda. More specifically, we discuss four aspects regarding RFID research issues and policy implications. First, we examine various competing concepts of RFID use by governments all over the world. Second, we categorize numerous applications of RFID technology through analyzing previous literature. Third, we try to figure out technological issues and governance problems that RFID technology faces today. Last, we draw key public issues and suggest future research agenda.

2. METHODOLOGY OF THE RFID LITERATURE REVIEW

2.1 A brief history of RFID technology

RFID technology was emerged as Frederick Hertz found existence of radio frequency during his experiment in 1886 (Wyld, 2005) and developed for the purpose of defense during the Second World War[4]. During 1970s and 1980s, the RFID system attracted plenty of scholars and innovators, so efforts to register patents progressed (Takahashi, 2004). Researchers like Charles Walton had registered a patent to use RFID. In the 1980s, many US and European companies recognized the importance of developing RFID technology and started to manufacture RFID tags. Soon scholars at MIT University opened an Auto-ID center to promote the use and implementation of RFID technology. But most of the scholars report that the first commercialization of RFID technology was done by Wal-Mart as they launched RFID based material identifying system in 2005 (Shahram and Manish 2005). Wal-Mart is now tracking merchandise including food, apparels, and electronic items with RFID technology in their supply chain.[5]. RFID technology is a brand new policy tool that can ensure high transparency, efficiency and effectiveness not only in industrial areas but also in government service delivery. Table 1 describes a brief history of how RFID technology was developed and diffused.

Table 1. A brief history of RFID technology

Date	Event
1886	The idea of using Radio Frequency to reflect waves from objects was started from Frederick Hertz's experiment.
1930–1940	American navy research laboratories developed a system known as IFF (Identify Friend or Foe).
1940–1950	The first application of RFID consisted of identifying allied or enemy planes during WW2 through the use of IFF system.
1973	Charles Walton, a former IBM researcher registered patent using RFID technology, a radio-operated door lock.
1980–1990	Many US and European companies started to manufacture RFID tags.
2003	The Auto-ID center for MIT became EPC global, an organization whose objective is to promote the use and adoption of RFID technology.
2005	Wal-Mart launched and RFID pilot.

2.2 Research design for a systematic review

We searched online data base and expert based information to identify RFID publications between 2003 and 2015. We categorized RFID applications and analyzed issues and concerns that RFID faces today by systematically reviewing published literature. We have collected literature we use for systematic review from two different resources. First, most of the studies are found by searching the e-database. We could access electronic databases, such as Google Scholar, World Web of Science (WWS), Proquest Central, and Science Direct through Seoul National University's main library homepage. We had set 'RFID technology', 'RFID government,' 'RFID application, and 'RFID issue' as keywords for searching literature. We found most of the research through this method of searching. The second method we used for collecting data was having discussions with experts. To do this, we first made a list of experts who specialize in IT technology, Science technology, and public administration. Five experts agreed to help us and recommended some research papers that were known for their fluent flow of logic and plentiful contents. We chose relevant research papers from among experts' recommendations. In sum we had used previous literature collected from two methods we discussed above, searching e-database and asking experts, as our resource of searching.

[Figure 1] shows analytical frame that we use for this study. We have determined the literature for systematic review according to three stages shown on the flow chart. First, the original total number of studies we have found from the e-database was 4260. Also 185 research papers were found from experts' recommendations and previous public papers. A total of 4,445 studies were chosen through the first stage. Second, we excluded 4,121 following general eligibility criteria by screening title and abstract. More specifically, we excluded RFID studies only with one of the following criteria: 1) studies focusing on private sector; 2) studies without considering how public sector implemented RFID technology; 3) studies that did not discuss any social scientific implications; and 4) studies that only deal with RFID technology from pure scientific and engineering points of view. In sum, we included only 324 papers that discussed RFID issues and their implications in public sector. Third, we removed further 213 studies too much focusing on private sector or RFID technology itself, rather on its applications in our society and social scientific implications. Finally, 111 articles were chosen for our systematic review.

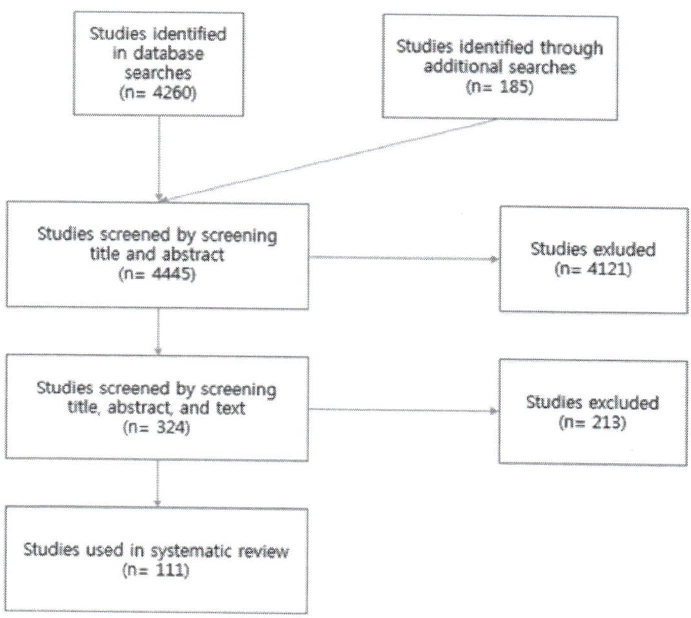

Figure. 1. Analytical Frame

[Figure 2] below showed descriptive statistics of collected literatures by published year. It shows 22 studies were published in 2007 among 111 literatures. As we already described above in history of RFID section, the popularization and commercialization of RFID technology was started in 2005 with Wal-Mart's adoption. It seems that after Wal-Mart's innovative footsteps hit the world, many scholars were started to recognize the potential of new technology and tried to understand and develop RFID technology. Besides some governments from all over the world implemented new way of public service delivery using RFID technology. Consequently, 49 literatures were publishedbetween 2006 and 2008 and it forms almost 45 % of our collected studies

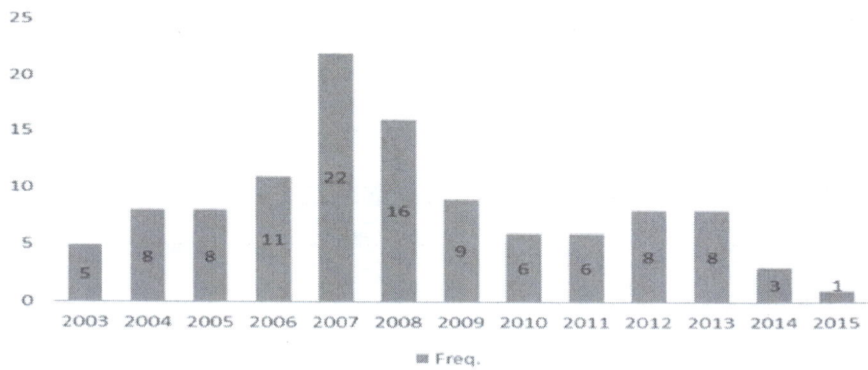

Figure. 2 . Descriptive Statistics of Literatures by year

For this study, we categorized governments' way of using RFID technology in 6 areas; Agriculture and Livestock, Defense and Security, Environmental Applications, Healthcare and Welfare, Identification, and Transportation. [Figure 3] shows descriptive statistics of collected literatures categorized by applications. We categorized studies that did not focus on specific sector and analyze and introduce RFID technology from the general perspectives as 'RFID general'. 'RFID general' studies usually deal with various ways of using RFID technology in diverse sectors simultaneously. As we can see from [Fig. 3], RFID general area had 42 papers. That means still lots of RFID studies could not be fully specialized and remained in status of generally introducing RFID technology. Identification sector scores secondly highest number of published literatures among areas. This result seems natural because e-ID card or e-Passport have most powerful force that can hurt privacy, one of the most serious and notorious issues that RFID technology face today

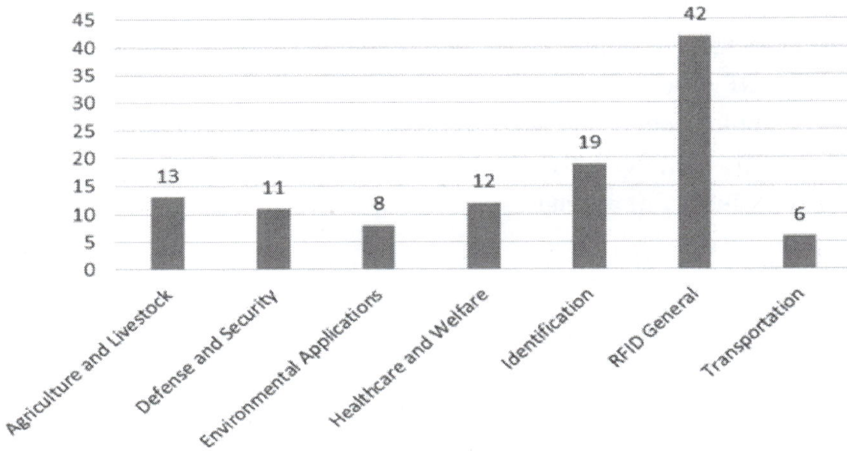

Figure. 3. Descriptive Statistics of Literatures by applications

2.3 Key applied areas of RFID

2.3.1 Defense and security

As we show in [Table 1], the history of RFID technology was started from the need for ensuring national security. Almost 60 years have passed since US army developed RFID based identification system to identify allies and enemies, RFID technology is still used for protecting people. For instance, Weinstein (2005) and Konsysnki & Smith (2003) reported how the US Army and Navy implement RFID technology in cargo containers to identify materials. The US Army and Navy implement RFID not only to identify US troops' own weapons and containers but also to identify enemies in battle (Tien 2004)[6]. RFID systems are also important in terms of airport and port security. After the 9/11 terror attack on the United States, President George W. Bush let all the airport and port in US adopt identification systems based on RFID technology to protect its nation from additional terrorist attacks (Werb and Sereiko 2002)[7]. In 2012, the Taiwanese government decided to implement an RFID based e-Seal system to increase security and efficiency (Tsai and Huang 2012). In addition, RFID technology can be used effectively in prison management[8] and child protection. In some countries like Japan and Republic of Korea, the RFID tag is implemented in child protection monitoring (Table 2).[9]

Table 2. RFID application in Defense and Security

Application	Author	Year	Case	Country
Army and Navy	Weinstein	2005	US Navy embedded RFID into cargo container	US
	Konsynski and Smith	2003	US Army track containers of material	US
	Tien	2004	US Army using RFID for tracking its army in Iraq	US
	Anon	2002	US Army piloted 4 projects using RFID	US
Airport and port Security	Werb and Sereiko	2002	2 RFID programs to prevent terrorist's attack	US
	Smith and Konsynski Zhang	2003 2013	New York city government project using RFID, *CargoMate*	New York City, US
	Tsai and Huang	2012	Kaohsiung port adopt RFID system for port security	Taiwan
Prison management And Child protection	Kim	2008	Calpatria prison issued RFID embedded bracelet to its inmates	US
	Nicholas	2008	LA County prison management	US
	Ema and Fujigaki	2011	Let parents know the exact time of a child's arrival and departure time	Japan
	Yonhap news	2013	Child monitoring through RFID tag in beach	ROK

2.3.2 Identification

Electronic passports like 'e-passports' were adopted electrically after the 9/11 attack. After the terrible tragedy broke heart of United States, the American government became aware of the importance of checking VISAs and passports correctly. The US Department of State soon let people who wanted to enter US to use RFID tag embedded electronic passports instead of traditional barcode based passports[10]. The European Union also endorsed the inclusion of biological information in e-passports. The EU Justice and Home Affairs Council decided to include fingerprints as a second mandatory identifier on passports in 2004.[11] In addition, RFID can be used in e-ID cards in various countries. For example, in the United Kingdom, Prime Minister Tony Blair and his Labor Party convinced the nation to adopt biometrically-enhanced national identification cards (Ezovski and Watkins 2007). Tony Blair's administration announced its will to implement RFID tag embedded national identification card in late 2004. China is another case where the e-ID card is used today. As a matter of fact, China is the country where e-ID card is widely and largely adopted today. The Beijing Olympics held in 2008 lit the fuse of adoption. The largest smart card project was implemented as a part of preparing the most prominent international sports event. In 2008, the Chinese government supplied 1.2 billion dollars of RFID readers and 2.25 billion dollars of RFID embedded smart cards to citizens. This made China the world's largest market for RFID (Kovavisaruch and Suntharasaj 2007) (Table 3).

Table 3. RFID application in identification

Application	Paper	Year	Case	Country
e-passport	Meingast et al.	2007	US Visit Visa Waiver Program(VWP) focusing on why US government adopted RFID based e-passport	US
	Lorenc	2007	History of e-passport	US
	Ramos et al.	2009	Disadvantage of using e-passport from privacy perspective	US
	Kim	2008	Pros and cons of ROK government using RFID embedded e-passport	ROK
	Hossain	2014	How Australia adopt and implement e-passport	Australia
e-ID	Ezovski & Watkins	2007	e-ID used in United Kingdom	UK
	Kovavisaruch & Suntahrasaj	2007	Brief history of Chinese e-ID card	China

2.3.3 Environmental applications

RFID technology can be widely applied in environmental applications. Adopting an RFID system in waste management is the most prominent way of using RFID to ensure efficient, eco-friendly waste management among lots of countries in the world. PAYT (Pay-As-You-Throw) program done by European Union (EU) is the pioneer of this field. PAYT is an RFID based waste pricing model that allows each individuals or each household to pay for the tag along with the total amount of waste they throw. Since each household and individual has a waste box in which RFID tags are embedded, the exact volume of waste can be calculated. In Europe this incentive based system has been proven to be a powerful policy tool for reducing the total amount of waste and for encouraging recycling (Schindler et al. 2012). Similar systems are broadly implemented in US (Ransford et al. 2012). In South Korea, the ministry of environment introduced it to industry and urged them to use an RFID based waste management system, especially in medical waste management. RFID technology is implemented in waste management in developed and developing countries, but the purpose of adoption is somewhat different from Europe to the US. India, the second-most populated country in the world, has adopted RFID technology to cope with the rapid increase of volume and types of waste (Infotech 2013). Similarly in 2010, what China faced were the World Expo and huge amounts of construction waste that comprised 30 % to 40 % of the total urban waste. Shanghai was chosen for a pilot project using an RFID based waste management system. All the waste dumping trucks had an embed RFID tag and volume of waste they carry was checked by the local government (Ruan and Hu 2011). Another interesting case of environmental application emerges from South Korea. The South Korean government operates U-Street Trees Systems through which the exact location and status of street trees can be monitored. Information about location and status of street trees are collected by an RFID tag that is attached to each tree is saved in a web information system, so trees can be managed effectively. Kim et al. (2006) claim that this web based information system could manage information remotely with an interactive system (Table 4).

2.3.4 Transportation

Public transportation is another popular sector for RFID technology applications. RFID based electronic toll collection technology is one of the oldest and widespread RFID implementation (Ulatowski 2007). As soon as an RFID tag embedded car arrives at a toll booth, the RFID reader scans and reads the information that the RFID tag contains. The driver will pay debits according to the price that electronic reader suggests. In the US, electronic toll collection is thought as efficient and effective method that eliminates long lines of traffic at toll booth (Ulatowski 2007). RFID based toll collection is also adopted in criminal cases because it enables prosecutors to identify the exact location of the criminal's car (Smith 2006). In South Korea, the Korean government has set credit card-linked electronic toll collection system called 'Hypass' especially for collecting transportation tolls on express ways. If an RFID tag is embedded on their cars, drivers can pass the tollbooth without stopping the car because RFID reader scan the data immediately and handle the whole payment process in about 5 s (Kim 2008). Hong Kong launched similar public transportation toll collection system in 1997 and the 'Octopus Card' is now internationally famous for its convenience. This system is able to handle 10 million transactions per day and includes all modes of public transport (Kovavisaruch and Suntharasaj 2007). South Korea has set credit card-linked electronic toll collection system called 'Hypass' especially for collecting transportation tolls on express ways. RFID technology is also implemented in railroad toll collection in India, where railroads are the most widely used form of public transportation. If an RFID tag is embedded on their cars, drivers can pass the tollbooth without stopping the car because RFID reader scan the data immediately and handle the whole payment process in about 5 s (Kim, 2008). In addition, RFID has been used as a critical technology to promote efficiency and transparency for public transportation system in developing countries. For instance, the Mexican government runs "Creating Traffic Knowledge in Mexico: Applying RFID to prevent vandalism" and one of the purposes of this innovative project is to develop a transportation information system to acquire more subtle data necessary for government decision making (Prado et al. 2010). Analogous to Mexican case, in Bangladesh where BRTA (Bangladesh Road Transport Authority) was started in 2003, the technology is operated mainly for control and supervision of the road transport systems (Hossain et al. 2009). RFID technology is also implemented in railroad toll collection in India, where railroads are the most widely used form of public transportation (Table 5).

Table 4. RFID applications in defense and security

Application	Paper	Year	Case	Country
Waste Management	Kang et al.	2012	Ministry of Environment in the ROK introduced RFID system in the medical waste management in 2006	ROK
	Infotech	2013	India adopted RFID technology in order to deal with rapid increase in volume and types of waste	India
	Ruan and Hu	2011	Shanghai city government started RFID using waste management to prepare Shanghai expo 2011	China
	WWICS	2008	How US government using RIFD in waste management	US
	Schindler et al.	2012	Cases of using RFID in waste management in EU member country	EU
	Ransford et al.	2012	Waste management operating system broadly used within US	US
Street tree Management	Kim et al.	2006	ROF government using RFID technology to manage street tree condition	ROK

Table 5. RFID application in transportation

Application	Paper	Year	Case	Country
Public Transportation	Ulatowski	2007	Electronic toll collection	US
	Kovavisaruch and Suntharasaj	2007	Well-known success smart card case in Hong Kong	Hong Kong
	Kovavisaruch and Suntharasaj	2007	Largest smart card transport system in the world	China
	Pransanth and Soman	2009	RFID based transportation system especially for railroad toll collection in India	India
	Kim	2008	Hypass case used in highway toll collection in ROK	ROK
Managing Road condition	Prado et al.	2013	How Mexican government using data collected from RFID system in decision making	Mexico
	Hossain et al.	2012	Monitor and control the road system	Bangladesh

2.3.5 Healthcare and welfare

RFID enables hospitals to manage their equipment more easily and save expenses in public health areas[12]. The US government agencies like FDA have also already used RFID tag in monitoring drug industry[13]. Since American hospitals handle almost 4,000 medicines per day, medication errors can be easily occurred. With strong government support, public hospitals in Taiwan have actively adopted innovation of RFID (Kuo and Chen 2008)[14]. Even though it is not yet commercialized, an RFID identification system[15] for the visually impaired people is being developed by engineers in Pakistan with the support of the Pakistani government (Murad et al. 2011) (Table 6).

Table 6. RFID application in healthcare and welfare

Application	Author	Year	Case	Country
Managing Public Hospital	Kuo and Chen	2008	Many hospitals are actively involved RFID system in managing hospital with the support of the government	Taiwan
Pharmacy	Wyld Thuemmler et al. Romero and Lefebvre	2008 2007 2013	U.S. government and FDA recommended that pharmaceutical industries move to implement RFID tag to prevent counterfeit dugs by 2007	US
	Skinar	2005	Florida state government imposed fine to drug suppliers when they did not adopt RFID tags	Florida State
Service for the impaired	Murad et al.	2011	RFID tag using service for visually impaired people was designed and implemented by Pakistan government	Pakistan
Infection Management	Nicholas	2008	Singapore government track visitors, patients and staff, to figure out who was the SARS virus bearer in 2003	Singapore
	Kuo et al.	2004	Taiwan hospital proceed RFID plans to track SARS patients	Taiwan

2.3.6 Agriculture and livestock

RFID technology can be an effective tool for securing food safety and managing agriculture and livestock. Another major advantage to this system is that animal disease tracking can be realized through innovative technology like RFID (Hossain and Quaddus 2009). With the government support, researchers have developed the Navigation System for Appropriate Pesticide Use as a basic system for risk management in agriculture (Nanseki et al. 2005). RFID technology in agriculture was first introduced by the European Union (EU) in the late 1990s and shortly thereafter many countries, such as Australia, Japan and South Korea, adopted the innovation. Among those countries, the Australian government was the most passionate in implementing RFID[16]. For instance, all the livestock in Australia have RFID embedded tags on their bodies immediately after they are born; information that enables farmers to identify each entity and its health status is registered in National Livestock Identification System (NLIS). RFID technology in Japan has been also adopted in agriculture especially to secure food safety and agricultural risk management that can occur by abusing pesticides (Nanseki et al. 2005, Sugahara 2009). The Japanese government planned to make a food traceability system by 2010 as a part of the "e-Japan" plan (Chen et al. 2008). The United States is another case that applies mandatory RFID based identification system in managing livestock. According to RFID Gazette (2006), the USDA is pushing for RFID tagging of cattle to make tracing of disease patterns easier. With the formation of National Institute for Animal Agriculture (NIAA) in 2002, the plan for setting the National Animal Identification System was started. What the US government fulfilled through this program was "to be able to identify all animals and premises that have had contact with a foreign or domestic animal disease of concern within 48 h" (Wyld, 2005) because "the sooner animal health officials can identify infected and exposed animals and premises, the sooner they can contain the disease and stop its spread (USDA-APHIS 2005)" (Table 7).

Table 7. RFID application in Agriculture and Livestock

Application	Paper	Year	Case	Country
Agricultural risk management	Nanseki et al.	2005	Navigation system for Appropriate Pesticide Use was developed as a system for agricultural risk management	Japan
	Sugahara	2009		
Livestock Management and Disease Tracking	Trevarthen and Michael	2007	Cochrane Dairy Farm let their cows in herd have National Livestock Identification System	Australia
	Hossain	2009	How Australian farms adopt and implement RFID technology	Australia
	Wyld	2008	US began National Animal Identification System in 2002	US

2.3.7 Public policy issues from RFID diffusion

RFID applications and diffusion generate complex policy and governance problems. We address public policy issues such as technological gap and uncertainty of expecting potential benefits and costs from a rapid and massive RFID diffusion. Uneasy governing issues in transparency, digital identification and power distribution are arising from inappropriate RFID applications. We discuss governance issues such as corruption, privacy problem, and digital monopoly and literacy in the following.

2.3.8 Technological concerns

Technology is not still enough to satisfy all the elements that RFID is trying to perform various operational mechanisms. RFID technology deficiencies inevitably occur with the application of technology because there is niche space still left. For instance, RFID technology does not have a unified frequency standard yet. Since there are no internationally agreed upon frequencies for RFID operations, permitted scanner/reader powers also differ between countries. There are still significant differences between the frequencies from the EU and the USA (Hossain et al., 2009). In addition, Reichenhach (2008) pointed out the lack of storage capacity. In the EU, where RFID based waste management is common, there are technological barriers like a shortage of storage capacity[17]. Ema and Fujigaki (2011) draw implications from a child monitoring case done in Japan that being informed of children's exact location cannot guarantee their actual safety, but RFID tags often lead to that cherished illusion. Vining (2005) warned about another possibility of niche space. According to his study about port security in the US, stealing goods without damaging RFID tag is possible because at ports, the container can be drilled into and contents can be removed. The RFID tag does not have to endure any damage through this whole process. In the US, as a response to continued pressure from various stakeholders, the US government even adopted the 'Faraday cage' for privacy protection[18] (Table 8).

2.3.9 Uncertain cost-benefit effectiveness

RFID defenders emphasize that RFID technology can guarantee effectiveness and efficiency at a very cheap price. There is, however, substantial evidence to show RFID can generate unexpected costs[19]. In reality, the RFID tag is much more expensive than a barcode, which was very popular in identifying materials before the rise of RFID technology (Becker 2004). Purchasing RFID devices, hardware, and tags is not sufficient to drive system relevantly. To guarantee a better quality of service, the RFID system needs more additional things such as "circular process mechanism, the richness of consultant, project manager, programmers and plentiful project labors" (Kuo and Chen 2008). These elements for a better RFID performance may involve considerable costs. Kuo and Chen (2008) reported that RFID technology consumers and government should pay the extra hidden cost in the healthcare industry (Table 9).

Table 8. Technological Issues

Issue	Paper	Year	Case	Country
Cryptography	Laurie	2007	Cryptography is not a perfect technique for protecting information saved in RFID tag so it can be easily attacked by the hackers	US
Standard	Hossain et al.	2009	No internationally agreed frequencies for RFID operations	
Storage Capacity	Reichenbach	2008	There is shortage of capacity for source-separated waste in household	Europe
NicheSpace	Ema and Fujigaki	2011	Informing children's exact location cannot guarantee their actual safety	Japan
	Vining	2005	In port, the container can be drilled into and contents can be removed without disturbing the RFID tag	US
	Ezovski et al.	2007	Without any additional protections, the Faraday cage is not safe enough	US

2.3.10 Dubious transparency and corruption

RFID technology is expected to increase transparency and monitor corruption. However, RFID technology cannot ensure a high level of transparency than expected. As a matter of fact, RFID tags can be cloned and manipulated quite easily, and this kind of tag corruption can occur at every stage of RFID implementation. There are various examples to show an inappropriate use of RFID technology. For instance, Armknecht et al. (2010a, b) warned the possibility of tag corruption. Lee et al. (2012) pointed out reader corruption of the RFID technology. An existing security model mainly focuses on the possibility of tag corruption, but reader corruption can hurt consumers' privacy as seriously as tag corruption can. Jules (2006) reported one of the tag corruption cases that observed in United States. One of the staff members who worked in a Dupyu store, an unscrupulous retailer, attached a cloned tag to counterfeit drugs. Avoine et al. (2010) argued that internet based databases can also be directly attacked and emphasized the possibility of reader corruption. There are also unethical behaviors to avoid RFID monitoring process. In the EU where an RFID based waste management system is aggressively implemented, some people disposed waste that came from their house at work places in order to avoid exact calculation through the RFID system. Not only this, Bilitewski (2008) reported that some conscienceless people are burning or transferring waste outside instead of throwing it into their RFID tag attached garbage can (Table 10).

Table 9. Cost-benefit effectiveness issues

Issue	Paper	Year	Case
Cost	Becker	2004	RFID tag is much more expensive than barcode
Hidden Cost	Kuo and Chen	2008	There is still hidden cost for highly qualified RFID system
Privacy	Jensen	2008	Active tag, which is more safe for securing privacy than passive tag is more expensive than passive tag

2.3.11 Privacy issues

One of the most serious issues that RFID technology faces today is whether RFID technology is secure enough to protect privacy. Privacy is the most important concern RFID users have to deal with (Perakslis and Wolk 2005). RFID tag embedded chips often contain important personal information and usually this kind of private information can hurt one's privacy seriously if leaked. To prevent leakage of private information, engineers developed cryptography, but there remains criticism[20]. The reason why these sorts of privacy concerns arise is because of the lack of security protection capacity of modern RFID technology. As we discussed above in the technological issues section, RFID technology today is not developed to secure perfect privacy. The technology itself has lots of deficiencies and people are smart enough to find niche spaces that can destroy the RFID security process. RFID itself can involve not only various hidden costs[21] but also induces a serious privacy problem[22]. However, despite these possibilities of attacks on privacy, there are lots of stakeholders and scholars who advocate the potential benefits of RFID. They claim that tracking and profiling consumers is solely for implementing RFID chips more effectively. Eaward Rerisi, one of the producers of early implementation of RFID technology argued that, "An RFID reader can read the number on a tag, but without knowing what the number means, there is no way to access personal information. The idea that the tags can be read by just anybody—that's pretty impossible" (Murray 2003) (Table 11).

Table 10. Transparency issues

Issue	Paper	Year	Case	Country
Tag Corruption	Armknecht et al.	2010	Tag corruption can be occurred at every stage of RFID implementation	
	Jules	2006	Dupyu staff attach cloned tags to counterfeit drugs	US
Reader Corruption	Lee et al.	2012	Although reader corruption can cause serious privacy attack, most of the scholars do not consider it as security problem.	
Cheating	Reichenbach	2008	Waste being disposed of at work places	EU
	Bilitewski	2008	Burning waste or transferring it to outside	EU

Table 11. Privacy issues

Paper	Year	Category	Contents
Hwang et al.	2009	Cloning	The attacker can read the tag and then clone the tag by writing all the obtained data into other
		Eavesdropping	The attacker surreptitiously listens to all the communications between the reader and the tag
		Replay attack	The attacker repeats or delays the same message when valid data are transmitted
		Denial of Service	The attacker can send massive message to RFID system and attempt to crash the RFID system
		Forward Security	The attacker can compromise a tag and obtain its current relation date
		Tag tracing	The tag always broadcasts a fixed serial number to somewhere nearby the reader; therefore, the adversary can identify a fixed serial number of the tag from different locations or transaction records
		Individual data privacy	The hacker can know what items the consumer bought from the store or what books the consumer borrowed from the library
		Data forging	The attacker can modify the dates, items, and prices and then cause great loss if the tag can store extra data
Numann and Hogben	2008	Skimming	The attacker opens a clan-destine connection to the chip and gains access to the data
		Eavesdropping	The attacker intercepts the communication between the chip and an authorized reader
		Location tracking	The attacker generates person or card-specific movement profiles.

2.3.12 Unequal power and digital literacy

Unequal distribution of RFID technology can generate unequal distribution of various resources such as information and digital literacy. Especially in developing countries, the combination of unbalanced power distribution between stakeholders and a low level of digital literacy can cause serious problems. Ketprom et al. (2007) emphasized that in developing countries like Thailand[23], governments should provide education and training on how to use brand new technology to poor farmers whose digital literacy remains relatively low. But poor farmers in Thailand are not the only stakeholders who are suffering from a lack of digital literacy. In Bangladesh, where RFID toll collection is common, traffic policies have no interest in using RFID technology for managing public transportation systems. Rather, they prefer traditional ways of toll collecting to information based technology (Hossain et al., 2009). Prasanth et al. (2009) found that the lack of digital literacy among the

Indian people hampered an effective process of railroad toll collection in India. Another problem developing countries face is an unbalanced power distribution due to lack of democratic value embedded governance. Chen et al. (2008) criticized the Taiwanese government because it monopolizes most of the information collected by RFID technology. As we already discussed above, when RFID tag scanned, information saved in RFID tag is scanned by reader and then transmitted to an internet based database. If that data were available to the public, individuals and industry could make more reasonable decisions by analyzing them. We find another unbalanced power distribution case in China's waste management system. According to Ruan and Hu (2011), the Chinese government benefits most from the RFID system[24] (Table 12).

Table 12. Power distribution and digital literacy issues

Issue	Paper	Year	Case	Country
Digital Literacy	Ketprom et al.	2007	RFID technology is widening the gap between poor and rich farmers	Thailand
	Hossain et al.	2009	Traffic policies are not interested in using RFID technology	Bangladesh
Power Distribution	Ruan and Hu	2011	Price of construction waste transportation is set by government, so other stakeholders have no choice	China
	Chen et al.	2008	Consumer can get only limited information	Taiwan

3. DISCUSSION AND CONCLUSION

We found, relying on a systematic review from 111 RFID studies, six key areas of RFID applications. Specifically in the defense and security section, we addressed how military and airports/ports manage RFID systems to ensure security. We also found that RFID is effectively implemented in prison management and child protection programs. Numerous governments have introduced RFID identification tools such as e-passport and e-ID. RFID systems for waste management and street tree management are widely used from rich to poor countries. In healthcare and welfare delivery, RFID based smart cards have turned out to be very efficient. RFID is now being used to monitor counterfeit drugs. RFID has been applied to delivering service for the impaired and to trace infection. However, despite potential benefits from RFID applications, various unexpected problems arise. RFID can still involve technological deficiencies, especially in securing cryptography techniques, international standards of frequency, and storage capacity. RFID technology is not still enough to be efficient and effective in some areas (Becker 2004, Jensen et al. 2007). Tag and reader corruption can hurt transparency and security. Privacy issues are still the most serious issues that RFID faces today (Naumann and Hogben 2008). RFID itself can generate new unequal digital literacy and power distribution, especially in developing countries such as Thailand and Bangladesh. Even the most latest innovative technologies, like RFID, do not have perfect answers to securing efficiency, effectiveness, convenience, and transparency. Rather, RFID technology itself creates unexpected problems. It should be noted that democratic governance and trust is still important to technological innovation and policy issues arising from a rapid RFID diffusion.

Our systematic review is incomplete to discuss all of the RFID issues from technology, market and management, e-government, and legal aspects. Further research on RFID diffusion and impact include not only various theoretical issues of but also legal and managerial problems. For instance, both qualitative and quantitative research is required to explore what factors are critical to adopt and implement new RFID technology in terms of governance and digital literacy. Both micro and macro approaches with massive data are also required to identify how RFID improve not only organizational performance in government agencies and various industry sectors but also quality of our life.

FOOTNOTES

1 For example, after serious attack by Osama Bin Laden on 9/11, the American government decided to implement an RFID tag embedded e-passport and VISA waiver program. The US government asked their member countries to implement e-passport by late 2005 and soon US

member countries like ROK and EU started to use e-passports. Currently, no one can enter to United States without an RFID tag based e-passport.

2 Especially in developing countries, governments usually adopt brand new IT technologies, but their low level of socio-economic infrastructures may prohibit the efficient operation of technology.

3 For instance, RFID applications may lack social virtues like trust, ethics, and democracy. It is essential to understanding how a rapid diffusion and massive applications of RFID generate conflict or harmony among human behaviors, digital literacy, institutional rules, and technology.

4 US Army and its allies could not only manage weapons and soldiers but also identify who was the enemy or not (Castro and Wamba, 2007). This whole project of developing an RFID based identifying system was known as IFF (Identify Friend of Foe).

5 See more various examples of RFID applications at the website (https://epic.org/privacy/rfid/)

6 In 2004, the US Army adopted RFID during the Iraq war to track Iraq troops. Not only these, the US Army piloted 4 projects using RFID; identifying material locations, weapons deteriorations, hazardous material tracking, and asset tracking (Anon 2002).

7 The New York City government also started an RFID e-seal pilot project in the New Jersey Port. Once RFID read and scan the tag, it can identify the contents of the container box. Also, the port of Tacoma and Seattle planned to adopt E-Seals, made of metal bolts with embedded RFID devices to ensure its security (Konsynski and Smith 2003).

8 Calpatria prison, located in Los Angeles, adopted a prisoner monitoring system using RFID chips in 2000 as the very first pilot using RFID in prison management in United States (Kim 2008). According to regulations of Calpatria prison, all the prison inmates were issued bracelets in which RFID chips were embedded. Since the pilot project at the Calpatria prison was successful, the local government let other prisons in Los Angeles adopt the innovative bracelet. The LA County prison started to use a brand new bracelet in response to the state government's order; it is reported that through its adoption the prison

could increase efficiency and effectiveness and decrease crimes occurred between inmates simultaneously (Nicholas 2008).

9 For instance, city governments in the Gifu and Osaka prefecture provided RFID tags that can be attached in students' schoolbags to public elementary schools (Ema and Fujigaki, 2011). Similarly, in Haewoondae beach, one of the most famous vacation spots in South Korea, Busan Metropolitan City provides for parents RFID embedded bracelet that enables tracking exact location of their child by a smart phone (http://news.naver.com/main/read.nhn?mode=LSD&mid=sec&sid1=102&oid=001&aid=0003353674).

10 This project began in 2002, but it took 3 years to fully implement for all 16 US passports (Meingast et al. 2007). The appearance of the e-passport is very similar to old passports, but woven into the paper of the passport, there is RFID tag that information about owner of the passport is included. Information about nationality, sex, age, and so on is scanned, as airport staff members scan the passport through RFID reader (Lorenc 2007). The US government did not stop at this point and adopted VISA Waiver Program. By 2005, the US member countries had to adopt RFID based e-passports and VISA Waiver Programs in order for their citizens to enter the United States because without e-passports, passengers could not be accepted at American points of entry. Today, the e-passport includes not only individual data but biological data, such as fingerprints (ICAO TAG MRTD/TWF 2004).

11 E-Government News, "EU Asks US for Time to Issue Biometric Passports". iDABC European e-Government Services, 1 April, 2005.

12 For instance, Wicks et al. (2006) reported that this RFID based hospital management system is very effective in reducing management costs because embedded RFID tags can track lost or hidden expensive equipment. Miller (1999) pointed out that the potential of RFID technology can be expanded in tracking the location of patients and controlling the drugs. Also, Chowdhury and Khosla (2007) argued that RFID technology can be effectively used not only in hospital equipment management but also in patient management.

13 According to statistics published by the Institute of Medicine (IOM), about 44,000 to 98,000 people die in the USA per year because of

improper drug administration (Kohn et al. 2000). To rectify this phenomenon, in 2004 the US government and US FDA recommended pharmaceutical industries to implement RFID tags to prevent the production of counterfeit drugs (Wyld, 2005). The Florida state government added legal regulations to this recommendation in 2006. If a pharmaceutical industry located in Florida does not attach RFID embedded tags on its products, it has to face substantial financial penalties (Skinar 2005).

14 For instance, RFID technology saved Singapore and Taiwan from SARS attack around 2003. Two public hospitals in Singapore in 2003 adopted RFID technology to track staff, patients, and visitors in order to trace people who carried the SARS virus. The RFID information database saved all the data collected from each individual's RFID tag for 21 days, which was thought to be long enough for expression of SARS virus (Nicholas 2008). A similar process was done in Taiwan too. During the SARS period, five hospitals, including Taipei Veteran's General Hospital, implemented RFID tags to track patients who had possibility contracted the SARS virus (Kuo et al. 2004) with strong government support.

15 One peculiar characteristic of this system as compared to other systems is that visually impaired people are both the reader carrier and the service beneficiaries, simultaneously. Generally in government service delivery using RFID technology, the service provider usually carries RFID tag readers and service beneficiaries usually act more passive roles by attaching RFID tags. But in this Pakistani case, the service beneficiaries can identify objects around them by operating RFID tag reader they have.

16 The Australian government passed legislation on mandatory use of RFID tags in the livestock industry, so Hossain and Quaddus (2014) categorized Australian case as a very rare and special adoption case. According to Trevarthen and Michael (2007)'s case study, one of the Australian farms where the RFID tag is implemented, farmers not only track the exact location of the cows but also check the condition, identify cows and even feed the new born cows automatically by using an RFID system.

17 Usually people throw waste in various places. They may throw it away at their houses or at work places, like an office. Since the RFID tag is only attached to a garbage can in house, it is impossible to track all types of waste throwing behaviors. Inevitably, this shortage of storage capacity leads to selective waste collection monitoring.

18 The faraday cage is an object in metal; proponents of this device argue that faraday cage can prevent hacker's attack because electronic devices are prevented from passing through the object (Ezovski et al. 2007). But speculation about stability of this technology still remains. Lorenc (2007) reported that if there is no additional technology, the faraday cage cannot preserve sensitive security.

19 As a matter of fact, there are two kinds of RFID tags, the passive tag and the active tag. These two tags provide owners with different benefits and liabilities. Active tags are implemented by a power source, such as small battery. Active tags are more efficient and safe in protecting privacy than passive tags, so sensitive organizations like military prefer active tags to passive tags. But ordinary consumers have less accessibility to active tags because active tags are much more expensive than passive ones (Jensen et al. 2007).

20 According to Laurie (2007), even without physically losing an RFID tag, private information can be stolen because what's inside RFID tags can be skimmed quite easily. If we can build a device that enables us to transmit an arbitrary number, invading the internet based database is theoretically possible.

21 For instance, Wal-Mart Stores Inc. and Procter & Gamble Co. in autumn 2003 planned a very interesting experiment to check the potential deficiencies of RFID technology. Customers of a Wal-Mart store in Broken Arrow, Oklahoma were secretly tracked through RFID tags while purchasing lipstick that Procter & Gamble Co. made (Barut et al. 2006).

22 For instance, Hwang et al. (2009) and Numann & Hogben (2009) categorized various kinds of privacy attack cases that can possibly occur. According Hwang et al. (2009), technological deficiency enables hackers to engage in cloning, eavesdropping, replay attack, denial of service, forward security, tag tracing, individual data privacy, and data forging. Specifically, the hacker can read the tag and then clone the tag (cloning),

surreptitiously listen to all the communications between and the tag (eavesdropping), repeat or delay the message (replay attack), send large amount of message to break down RFID system (denial of service), compromise a tag (forward security), trace the exact location of the tag (tag tracing), find out shopping trend of the consumer (individual data privacy), and modify information saved on an RFID tag (Data forging). In addition, Numann and Hogben (2008) categorized the privacy attacking cases more briefly. According to their research, the hacker can attack RFID tag in some ways. First, the attacker can open a connection to the chip and can steal the data inside (skimming). Second, the attacker can intercept the communication between tag and reader (eavesdropping). Last, the attacker can track the exact location of the tag or the person.

23 In Thailand, most of the farms are trying to adopt an RFID system in farm management, but RFID technology is widening the gap between poor and rich farmers. Poor farmers are usually less educated people who have hardly had any experience using digital technology like RFID. On the other hand, well-educated wealthy farmers face low entry barriers and easily adopt the technology. Rich farmers armed with innovative technology not only make enormous fortunes by increasing efficiency but also by replacing poor labors with RFID embedded devices. According to the Thailand Ministry of Labor (2006), most farm labors are afraid of being replaced. Unfortunately, this phenomenon eventually correlates to a serious gap between rich and poor.

24 The government has invested a huge amount of money to buy necessary devices such as readers, tags, hardware, and so on, but in the long run cost of management will decrease. On the other hand, the waste industry could carry a very heavy debt. The waste contractors have to deal with expensive RFID tag rental and as well as the cost of construction simultaneously. Although this situation is totally unfavorable for them, the waste management industry cannot resist to this policy because the Chinese government is the entity which has made the use of RFID policy and set prices. The industry has no other choice but consent.

ACKNOWLEDGMENTS

This paper was supported by the research grant of Seoul National University Foundation (Korea Institute of Public Affairs) in 2015.

AUTHORS' CONTRIBUTIONS

KJ and SL carried out a systematic literature review not only from not only technological point of view but also from social scientific point of view. Both authors read and approved the final manuscript.

REFERENCES

1. Armknecht F, Chen L, Sadeghi AR, Wachsmann C. Anonymous authentication for RFID systems. In: Radio Frequency Identification: Security and Privacy Issues. Berlin Heidelberg: Springer; 2010a. p. 158–75.View Article

2. Armknecht F, Sadeghi AR, Visconti I, Wachsmann C. On RFID privacy with mutual authentication and tag corruption. In: Applied Cryptography and Network Security. Berlin Heidelberg: Springer; 2010b. p. 493–510.View Article

3. Avoine G, Coisel I, Martin T. Time measurement threatens privacy-friendly RFID authentication protocols. In: Radio Frequency Identification: Security and Privacy Issues. Berlin Heidelberg: Springer; 2010. p. 138–57.View Article

4. Barut M, Brown R, Freund N, May J, Reinhart E. RFID and corporate responsibility: hidden costs in RFID implementation. Bus Soc Rev. 2006;111:287–303.View Article

5. Becker C. A new game of leapfrog? RFID is rapidly changing the product-tracking process. Some say the technology--once costs drop--could displace bar-coding. Modern Healthcare. 2004;34:38–40.

6. Bilitewski B. From traditional to modern fee systems. Waste Management. 2008;28(12):2760–6.View Article

7. Castro L, Wamba SF. An inside look at RFID technology. J Technol Manag Innov. 2007;2:128–41.

8. Chen RS, Chen CC, Yeh KC, Chen YC, Kuo CW. Using RFID technology in food produce traceability. WSEAS Transac Inform Sci Appli. 2008;5:1551–60.

9. Chowdhury B, Khosla R. RFID-based hospital real-time patient management system. In: Computer and Information Science, 2007. ICIS 2007. 6th IEEE/ACIS International Conference on. 2007. p. 363–8. IEEE.

10. Ema A, Fujigaki Y. How far can child surveillance go?: Assessing the parental perceptions of an RFID child monitoring system in Japan. Surveillance Soc. 2011;9:132–48.

11. Ezovski GM, Watkins SE. The electronic passport and the future of government-issued RFID-based identification. In: RFID, 2007. IEEE International Conference on. Grapevine, TX: IEEE; 2007. p. 15–22.View Article

12. Hossain MA. Exploring the Perceived Measures of Privacy: RFID in Public Applications. Aus J Inform Systems. 2014;18(2):133–48.

13. Hossain MA, Quaddus M. An adoption-diffusion model for RFID applications in Bangladesh. In: Computers and Information Technology, 2009. ICCIT'09. 12th International Conference on. 2009. p. 127–32. IEEE.View Article

14. Hossain MF, Sohel MK, Arefin AS. Designing and implementing RFID technology for vehicle tracking in Bangladesh. Dhaka, Bangladesh: National Conference on Communication and Information Security; 2009.

15. Hossain MA, Quaddus M. Developing and validating a hierarchical model of external responsiveness: A study on RFID technology. Inform Systems Frontiers. 2014;17(1):109–25.

16. Hwang MS, Wei CH, Lee CY. Privacy and security requirements for RFID applications. J Comput. 2009;20:55–60.

17. Jensen A, Cazier J, Dave D. The impact of government trust perception on privacy risk perceptions and consumer acceptance of residual RFID technologies. AMCIS 2007 Proceedings, 146. (2007).

18. Jensen AS, Cazier JA, Dave DS. Mitigating consumer perceptions of privacy and security risks with the use of residual RFID technologies through governmental trust. J Inform Syst Security. 2008;4:41–66.

19. Jules A. RFID security and privacy: A research survey. Selected Areas Commun IEEE J. 2006;24:381–94.View Article

20. Infotech. RFID based Waste Management System. 2013. http://www.slideshare.net/iaitoinfotech/rfid-in-waste-management-slide-share.

21. Ketprom U, Mitrpant C, Lowjun P. Closing digital gap on RFID usage for better farm management. In: Management of Engineering and Technology, Portland International Center for. Portland, OR: IEEE; 2007. p. 1748–55.

22. Kim JG. A divide-and-conquer technique for throughput enhancement of RFID anti-collision protocol. Communications Letters, IEEE. 2008;12:474–6.View Article

23. Kim EM, Pyeon MW, Kang MS, Park JS. A management system of street trees by using RFID. In: Web and Wireless Geographical Information Systems. Berlin Heidelberg: Springer; 2006. p. 66–75.View Article

24. Konsynski B, Smith HA. Developments in practice x: Radio frequency identification (rfid)-an internet for physical objects. Commun Assoc Inform Systems. 2003;12:19.

25. Kovavisaruch L, Suntharasaj P. Converging technology in society: opportunity for radio frequency identification (RFID) in Thailand's transportation system. In: Management of Engineering and Technology, Portland International Center for. Portland, OR: IEEE; 2007. p. 300–4.

26. Kuo CH, Chen HG. The critical issues about deploying RFID in healthcare industry by service perspective. In: Hawaii International Conference on System Sciences, Proceedings of the 41st Annual. Waikoloa, HI: IEEE; 2008. p. 111–1.

27. Kuo F, Lee Y, Tang CY. The Development of RFID in Healthcare in Taiwan. Bejing: ICEB; 2004. p. 340–5.

28. Laurie A. Practical attacks against RFID. Network Security. 2007;2007:4–7.

29. Lee K, Nieto JG, Boyd C. A state-aware RFID privacy model with reader corruption. In: Cyberspace Safety and Security. Berlin Heidelberg: Springer; 2012. p. 324–38.

30. Li S, Visich JK, Khumawala BM, Zhang C. Radio frequency identification technology: applications, technical challenges and strategies. Sensor Review. 2006;26:193–202.

31. Lorenc ML. Mark of the Beast: US Government Use of RFID in Government-Issued Documents. The Alb LJ Sci Tech. 2007;17:583.

32. Meingast M, King J, Mulligan DK. Embedded RFID and everyday things: A case study of the security and privacy risks of the US e-passport. In: RFID, 2007. IEEE International Conference on. Grapevine, TX: IEEE; 2007. p. 7–14.

33. Murad M, Rehman A, Shah AA, Ullah S, Fahad M, Yahya KM. RFAIDE—An RFID based navigation and object recognition assistant for visually impaired people. In: Emerging Technologies (ICET), 2011 7th International Conference on. Islamabad: IEEE; 2011. p. 1–4.

34. Murray C. Privacy concerns mount over retail use of RFID technology. EE Times. 2003;2:4–5.

35. Nanseki. A navigation system for appropriate pesticide use: design and implementation. Agricultural Information Research. 2005;14:207–26.

36. Naumann I, Hogben G. Privacy features of European eID card specifications. Network Security. 2008;8:9–13.

37. Perakslis C, Wolk R. Social acceptance of RFID as a biometric security method. In Technology and Society, 2005. Weapons and Wires: Prevention and Safety in a Time of Fear. ISTAS 2005. Proceedings. 2005 International Symposium on (pp. 79-87). IEEE. New York, New York; 2005.

38. Prado JAD, Monterrey IC, Prado FED. Creating Traffic Knowledge System in Mexico: Applying RFID to Prevent the Vandalism. The 15th International Business Information Management Association Conference. 2010. p. 2042–50.

39. Prasanth V, Hari PR, Soman KP. Ticketing Solutions for Indian Railways Using RFID Technology. In: Advances in Computing, Control, & Telecommunication Technologies, 2009. ACT'09. Trivandrum, Kerala: International Conference on. 2009. p. 217–9. IEEE.

40. Ransford B, Sorber J, Fu K. Mementos: system support for long-running computation on RFID-scale devices. Acm Sigplan Notices. 2012;47:159–70.

41. Reichenbach J. Status and prospects of pay-as-you-throw in Europe–A review of pilot research and implementation studies. Waste Management. 2008;28:2809–14.

42. Romero A, Lefebvre E. Gaining Deeper Insights into RFID Adoption in Hospital Pharmacies. World. 2013;3:164–75.

43. Ruan T, Hu H. Application of an RFID-based system for construction waste transport: a case in Shanghai. In: Computational Logistics. Berlin Heidelberg: Springer; 2011. p. 114–26.View Article

44. RFID Gazette. RFID hybrid tech: Combining GPS for location tracking. Retrieved December 5, 2006, from tec.html http://www.rfidgazette.org/2006/12/index.html.

45. Schindler R, Schmalbein N, Steltenkamp V, Cave J, Wens B, Anhalt A, et al. SMART TRASH – study on RFID tags and the recycling

industry, technical report. TR-1283-EC. Santa Monica, CA: RAND Europe Corporation; 2012.

46. Shahram M, Manish B. RFID Field Guide: Deploying radio frequency identification systems. New York: Prentice Hall; 2005.

47. Smith JE. You Can Run, But You Can't Hide: Protecting Privacy from Radio Frequency Identification Technology. NCJL Tech. 2006;8:249.

48. Sugahara K. Traceability system for agricultural productsbased on RFID and mobile technology. In: Computer and Computing Technologies in Agriculture II, vol. 3. US: Springer; 2009. p. 2293–301.

49. Tien L. RFID tags should track inventory, not people. RCR Wireless News, 2004. http://www.rcrwireless.com/20040705/archived-articles/rfid-tags-should-track-inventory-not-people.

50. Takahashi. "The Father of RFID," Mercury News, www.siliconvalley.com (June 7, 2004).

51. Trevarthen A, Michael K. Beyond mere compliance of RFID regulations by the farming community: a case study of the Cochrane dairy farm. In: Management of Mobile Business, 2007. ICMB 2007. International Conference on the. Toronto, Canada: IEEE; 2007. p. 8–8.

52. Tsai FMC, Huang CM. Cost-Benefit Analysis of Implementing RFID System in Port of Kaohsiung. Procedia-Social Behav Sci. 2012;57:40–6.

53. Ulatowski LM. Recent developments in RFID technology: weighing utility against potential privacy concerns. ISJLP. 2007;3:623.

54. USDA APHIS. National Animal Identification System: Animal Identification Number(AIN). Retrieved from the web on June 8, 2005. Available at https://www.aphis.usda.gov/traceability/downloads/NAIS_overview_report.pdf.

55. Vining J. RFID Alone Can't Resolve Cargo Container Security Issues. 2005.

56. Weinstein R. RFID: a technical overview and its application to the enterprise. IT professional. 2005;7:27–33.View Article

57. Werb J, Sereiko P. More than just tracking. Frontline Solutions. 2002;3:11–42.

58. Wicks AM, Visich JK, Li S. Radio frequency identification applications in hospital environments. Hosp Top. 2006;84:3–9.View Article

59. Wyld DC. Delta airlines Tags Baggage with RFID. Re:ID Magazine. 2005;1:63–34.

60. Wyld DC. RFID: The right frequency for government. Washington DC: IBM Center for the Business of Government. IBM Center; 2005.

61. Wyld DC. Death sticks and taxes: RFID tagging of cigarettes. Int J Retail Distribution Manag. 2008;36:571–82.

62. Yonhap news. 2013. http://news.naver.com/main/read.nhn?.mode=LSD&mid=sec&sid1=102&oid=001&aid=0003353674.

63. Zhang R. A transportation security system applying RFID and GPS. J Ind Engr Manag. 2013;6(1):163–74.

Index

A
Agriculture, 118, 260, 267, 268, 286
Anti-theft system, 157

B
Bernoulli Random Graph, 75
Business Models, 1, 7, 14, 16

C
Chain Risk Management, 17
Check-Out Point, 167
communication technology, 50, 174, 180, 189
Complete Weighted Graph, 79
computational architectures, 174, 185
connected vehicle technology, 50, 61
cyber physical systems, 174

D
DCS Scenario, 90
Down syndrome, 228, 229, 230, 231, 232, 237, 242, 244, 247, 249

H
Healthcare, 46, 118, 253, 260, 266, 281, 284
Hybrid UHF/UWB system, 123

I
Integrated circuit, 127
interaction, 50, 175, 215, 223, 228, 229, 234, 237, 239, 243, 244, 245, 247, 248, 251, 253
interference model, 72, 73, 75, 81, 83, 84, 85, 86, 87, 88, 89, 91, 93, 95

L
literacy, 228, 229, 230, 231, 234, 235, 238, 245, 247, 248, 249, 250, 256, 269, 273, 274, 275, 276
livestock, 103, 256, 267, 278

O
On-chip antennas, 125

Q
Quasi Unit Disk Graph, 77

R

Random Proximity Graph, 80
reader-to-reader collision, 71, 72, 73, 74, 76, 77, 78
Response speed, 26
RFID smart sensor network, 174, 204
RFID-systems, 5

S

smart city, 174, 175, 212, 219
Smart dressing rooms, 156
Smart mirror, 156

T

tangible user interface, 228, 250
Tangible User Interface, 228, 232, 235, 237, 247
track and trace system, 174, 184, 186, 209
Transportation, 40, 50, 66, 68, 69, 220, 260, 264

V

Vehicle Operational Control, 61
vehicle positioning, 49, 50, 51, 52, 53, 60, 67, 68

W

Warehousing, 148